PENGUIN BOOKS

THE MOON BOOK

Bevan M. French received an A.B. degree in geology from Dartmouth College in 1958, an M.S. from the California Institute of Technology in 1960, and a Ph.D. from Johns Hopkins University in 1964. That same year he joined the NASA Goddard Space Flight Center and remained there for eight years, studying ancient terrestrial meteorite craters. In 1968 he was a visiting professor at Dartmouth. Beginning in 1969, he studied the lunar rocks returned from the Apollo 11, 12, and 14 missions. He was also selected as one of a small group of scientists to study material returned by the Russian Luna-16 unmanned probe, and in 1971 and 1972 he participated in astronaut training trips with the Apollo 16 and 17 crews. In 1972 Dr. French became Program Director for Geochemistry at the National Science Foundation, administering the support of basic research in the field of earth sciences. In 1973 he participated in the discovery of an ancient Brazilian impact crater twenty-five miles in diameter and 150,000,000 years old. He returned to NASA in 1975 as Program Chief for Extraterrestrial Materials Research; in this present position he administers NASA's research program on lunar samples, meteorites, and cosmic dust. Dr. French has published more than thirty-five technical papers about natural iron minerals, chemical reactions in rocks, terrestrial meteorite craters, and lunar samples. He lives with his wife in Chevy Chase, Maryland; they have three children and two grandchildren. Dr. French spends his spare time cutting grass, composing and playing folk songs, and continuing his research on terrestrial meteorite craters.

THE

BOOK

BEVAN M. FRENCH

PENGUIN BOOKS

Penguin Books Ltd, Harmondsworth,
Middlesex, England
Penguin Books, 625 Madison Avenue,
New York, New York 10022, U.S.A.
Penguin Books Australia Ltd, Ringwood,
Victoria, Australia
Penguin Books Canada Ltd, 2801 John Street
Markham, Ontario, Canada L3R 1B4
Penguin Books (N.Z.) Ltd, 182–190 Wairau Road,
Auckland 10, New Zealand

First published 1977

Library of Congress Cataloging in Publication Data

French, Bevan M
The moon book.

Bibliography: p.
Includes index.
1. Moon. I. Title.
QB581.9.F73 559.9'1 76–45672
ISBN 0 14 00.4340 3

Printed in the United States of America by
Murray Printing Company, Westford, Massachusetts

Set in Linotype Caledonia

dedicated
to my mother

MRS. JOHN S. FRENCH

who never minded going on trips
to collect rocks

TABLE OF CONTENTS

LIST OF ILLUSTRATIONS

PHOTOGRAPHS

Following page 116

FIGURES

ACKNOWLEDGMENTS

Although men first landed on the moon almost eight years ago, the study of lunar samples and the analysis of the data returned by the instruments on the moon is still an active and growing area of science. Published scientific results include over 5,000 technical articles and cover many yards of bookshelf space. Boiling down this flood of information into a single book, identifying the major scientific discoveries, and presenting the information in a readable format, would have been impossible without help from many people both inside and outside of the Apollo Program.

Many scientists active in the lunar research program were generous with both their time and their data. I am especially grateful to all those who contributed photographs: E. C. T. Chao, U. S. Clanton, R. L. Fleischer, B. P. Glass, S. E. Haggerty, O. B. James, D. S. McKay, E. L. Roedder, G. Simmons, N. Toksoz, R. M. Walker, and D. F. Weill. I regret that we were not able to use all the beautiful art work that was offered. John S. Shelton kindly gave permission to reproduce from his book *Geology Illustrated* the photograph of Meteor Crater, Arizona, that is photo 24 of this book. Other illustrations were supplied by many helpful people in the NASA Headquarters Office of Public Affairs, Washington, D.C., especially A. L. Gaver, and at the National Space Science Data Center in Greenbelt, Maryland. Dr. W. S. Cameron at the Data Center helped track down several fine lunar photographs.

The manuscript could not have been prepared without the careful typing of Alice McKinney and the photographic work of Lois Hazell. George W. Wetherill's excellent article on the ages of lunar rocks (*Science*, vol. 173 [July 20, 1971], pp. 383–92) provided the title for chapter 8.

The final text owes a great deal to the care and persistence of several readers: William and Kathleen Childs, Sharon Childs Moore, John A. O'Keefe, Ed and Kathleen Roedder, and Paul Lowman. Additional effort by Susan Zuckerman at Penguin Books helped produce a shorter, tighter, and (I must admit) better book. A special debt is owed to Mary-Hill French, who successfully handled the conflicting functions of colleague, editor, and wife during the years since 1972, when this book first began to take shape.

All of us who have had anything to do with the Apollo Program owe a great deal to the American public. Their interest, curiosity, excitement, and support made possible the discoveries that I have written of here. In a real sense, the discoveries of the Apollo Program are their discoveries, and this is their book.

Bevan M. French
Chevy Chase, Maryland

THE MOON BOOK

CHAPTER

THE MOON ROCKS ARRIVE

Toward the end of the summer of 1969, a number of people did something that no human being in history had ever done before; they held some of the moon in their hands.

For all of mankind the important date was July 20, 1969, when Neil Armstrong and Edwin Aldrin landed their spacecraft *Eagle* on the moon. Television made it possible for more than a half a billion people to watch as the astronauts planted the American flag on the moon, walked on the lunar surface, set out instruments, and picked up rocks.

Among the viewers were several hundred men and women with a special interest—the scientists who had eagerly prepared to study the returned moon rocks. For them, the important date was July 25 when, at about noon, two boxes containing the Apollo 11 samples were delivered safely to the Lunar Receiving Laboratory at NASA's Manned Spacecraft Center in Houston, Texas.

The 21.7 kilograms (48 pounds) of lunar samples soon became the focus of a program of investigation and analysis that was unique in the history of science. During the previous eight years, while hundreds of thousands of people had labored to send men to the moon and to bring them back safely, hundreds of others had been planning how the samples would be stored, preserved, and analyzed. A Preliminary Examination Team of about 50 scientists had been selected to conduct the first brief studies, after which samples would be distributed to about 150 carefully chosen groups of Principal Investigators and their associates. These scientists would analyze the moon rocks using every method that had been applied or could be applied to terrestrial rocks, obtaining information that would illuminate the origin and history of the moon in much the same way that earlier studies had helped us understand the evolution of our own planet.

On July 25, 1969, the work began. The samples were stored in huge vacuum chambers. The chemists and geologists of the Preliminary Examination Team crowded around with cameras as the sealed containers were opened. All that was visible at first was the featureless black dust that coated rocks and sample bags alike. Eventually the samples were dusted off, examined through the thick glass windows, probed, and described (photo 1). Small bits were cut off for microscopic observation and chemical analysis. As the distribution of samples to the Principal Investigators began, excitement and curiosity about the moon rocks grew and spread. The world was, in effect, looking over the scientists' shoulders to see what the rocks would reveal about the moon.

Not even the scientists themselves knew what to expect. Since prehistoric times man had watched and wondered about the moon. Just prior to the Apollo missions, the Ranger and Orbiter spacecraft had radioed back tens of thousands of close-up pictures of the moon. Surveyor landers had already dug trenches in the lunar rubble and analyzed its chemistry. But despite all this effort, no one could be sure of what the unopened sample boxes actually contained.

This mixture of curiosity and uncertainty was nothing new in man's history. The lunar landings were only the latest episode in man's perennial attempt to answer questions about himself and his surroundings: Is man unique in the universe? Is he alone? What kind of universe does he live in? How did it form? What is its past? What is its future? Is there proof of a Creator?

From the beginning of time man has attacked these questions with all the technical and mental abilities he possessed. Art, literature, religion, and philosophy have always been preoccupied with these themes. Science, from its hesitant beginnings three thousand years ago to its rapid expansion in the last three centuries, has supplied man with more and more information about his physical surroundings. As scientific knowledge has grown, man has asked more specific questions and has developed a new confidence that such questions could be definitely answered: How and when did the earth and the solar system form? What are the other planets like? Do they have life? How are the earth and moon different? Did they form in different ways? Did the earth look like the moon when it first formed? How did it change? Did the moon ever have an atmosphere and life like the earth?

At first, progress in answering these questions was slow. One reason was that the accessible earth and the inaccessible moon and planets* were studied by two separate groups of scientists. The study of the earth has been the province of geology, and geologists were able to study the

* Technically the moon is a satellite, and not a planet, because it revolves around a planet (the earth) instead of around a star. However, it is convenient to include the moon with the planets when discussing the nature, origin, and evolution of the solar system. This convention is not strictly accurate, but, as at least one lunar scientist has pointed out, it results in simpler sentences.

earth's rocks and decipher their history without any reference to the other planets. Astronomers, for their part, looked *away* from the earth instead of down into it, observing the stars and planets through the earth's murky, shifting atmosphere and usually thinking of the earth only as a firm base on which to build bigger telescopes. But no matter how close the largest telescopes brought the planets, it was still not close enough to see what they were really like.

Furthermore, astronomy is one of the oldest sciences whereas the scientific study of the earth is little more than a century old. The great scientific revolution between 1500 and 1700, spurred by the work of Copernicus, Brahe, Kepler, Galileo, and Newton, had firmly established the basic nature and mechanics of the solar system. It took another two centuries before earth scientists enacted a similar revolution in the understanding of our own planet. Not until the twentieth century did scientists develop the tools and insights to demonstrate that the earth is an ancient, continuously changing world that formed about 4½ billion years ago and has been altered by wind, water, volcanic eruptions, and mountain-building ever since.

But for a long time the origin and early history of the earth remained a mystery. Evidence from meteorites indicates that the solar system (and presumably the earth) formed about 4.5 to 4.7 billion years ago, but no rocks older than about 3.5 billion years had ever been found on the earth's surface.*

The first 25 percent of earth's history is missing. The rocks that existed between 4½ and 3½ billion years ago have been eroded by wind and water or have been consumed in ancient episodes of volcanism and mountain-

* In 1971, while the study of lunar samples was at its height, a group of scientists from Oxford University, England, reported the discovery of rocks nearly 3.8 billion years old in western Greenland.

building. The destruction of these ancient rocks has removed the record of how the earth was formed, produced volcanoes, developed an iron core and a magnetic field, produced oceans and an atmosphere, and, most important, developed life. As the Space Age developed, geologists found themselves in the position of a detective who must reconstruct the childhood of a man from information about only his adult life.

In July of 1969, the returned moon rocks united geologists and astronomers. The moon was placed under a microscope through which geologists could study its rocks with the same concern that they had lavished on earth rocks. Astronomers, meanwhile, waited to see what the direct study of the moon might tell about the origin of the solar system while geologists hoped that in the rocks from Tranquillity Base they might find clues to the missing record of the earth's childhood.

Even after all the planning and expectation that the Apollo Program had generated, the actual physical contact with the moon seemed to produce a sense of unreality in everyone, scientists and laymen alike. "You might as well wish for the moon," had once expressed the ultimate in impossibility. Now the meaning had changed completely; the moon had suddenly become something which could be held in the palm of one's hand, could be weighed, and analyzed. The change took some getting used to.

I first held a lunar sample in late September, 1969. I was visiting a colleague, Professor Malcolm Brown, at the University of Durham in England a few days after the first of his Apollo 11 samples had arrived. Durham is an interesting old city with an eleventh-century cathedral, and I was struck by the contrast between the ancient buildings of the town and the modern, newly installed safe in Professor Brown's office. Opening the safe, he carefully extracted a glass vial about the size of my little finger and handed it to me. I took it automatically. It contained about a teaspoonful of what looked like coarse black sand. When I first realized that I was holding a piece of the moon, my reaction was, "It isn't possible." Eventually, after working with many other samples from different parts of the moon, I got used to the idea, but this initial feeling of marvelous unreality never vanished entirely.

The contents of the bottle were the color of crushed coal and seemed too dark to be part of the moon. But that was how the moon had to look at close range. Long before Apollo we knew that the moon, despite its brilliance in the night sky, is a very poor reflector, reflecting only about 7 percent of the sunlight that falls on it. I knew that the surface of the moon had to be this color, but it still seemed strange to bring back such black stuff from such a bright object.

Curiosity overcame awe, and I turned the vial over in my hand to examine the contents, producing a clear, light, tinkling sound as the fragments rolled against the glass. The black mass in the bottle resolved itself into individual fragments that ranged in size from small peas down to pinheads. With a magnifying glass I recognized different kinds of fragments. There were small dense rock particles in which tiny crystals gleamed. Other fragments were porous and slaggy looking, with hard, sharp edges and open holes; these looked as though they had been heated to the point where they had begun to melt and bubble. I could also see a few shining spheres of glass which I identified as the "glass beads" that I had read about in the newspaper.

The first lunar sample I examined was not a single rock, but a complicated mixture of crushed and melted fragments. The bottle contained "lunar soil" (see Chapter 7) from the layer of shattered and melted rubble that covers the surface of the Sea of Tranquillity. Most of the nearly 22 kilograms of samples brought back by Apollo 11 consisted of this complex fragmental material; the 20 larger rocks that had been collected were still being studied in Houston.

Even before I picked up the bottle of lunar soil in Durham, scientists had already begun to analyze the samples, using instruments far more sensitive than the human eye and ear. The study of the moon had suddenly taken on a new dimension, since having actual rocks to analyze made it possible for the first time to probe directly into the moon's past.

"Rocks remember," one famous geologist observed, meaning that their structure, their minerals, and even the arrangement of their atoms, preserve a record of their formation and change over millions of years. But to read this preserved record requires both great insight and fine instruments, and only in the last 50 years has our knowledge and technology reached the point where we could establish a fairly clear picture of the history of our own planet. Now these abilities were applied to the moon rocks, and information began to pour out, not over a period of years, but during the course of days and weeks.

On September 15, 1969, only two months after Apollo 11 was launched to the moon, a scientific press conference was held in Washington, D.C., to present the information that the moon rocks had yielded. The results of this long-planned, intense analytical effort would have astounded anyone accustomed to the more leisurely pace of scientific research in the past. In only two months, using tiny amounts of material gathered from one spot on the moon (a site in Mare Tranquillitatis, or the Sea of Tranquillity), scientists were able to settle forever many major questions about the moon's origin and history:

1 / The rocks from the Sea of Tranquillity are basalt, a rock formed by the cooling of molten lava. This discovery revealed that the moon was once hot enough to melt inside. Therefore, we knew that the moon can no longer be viewed

as a cold, primordial, unchanging body. It is a unique planetary body with a history of its own.

2 / The basalt lavas in the Sea of Tranquillity flowed out onto the lunar surface about 3.7 billion years ago and have been preserved unchanged since then. From this we deduced that the moon is at least 3.7 billion years old and may be even older.

3 / The lunar basalts are similar to basalt lavas found on the earth in places like Hawaii, Iceland, and the Columbia River Plateau in the northwestern United States. Both lunar and terrestrial lavas contain the same chemical elements—chiefly silicon, aluminum, and oxygen. About 95 percent of both lunar and terrestrial basalts is made up of the same minerals. Thus the moon seemed to have close chemical similarities to the earth.

4 / In many ways, lunar and terrestrial basalts differ distinctively. The lunar basalts contain no water (even the freshest earth basalts contain a small percentage of water). Because of this absence of water, the minerals in lunar basalts are fresh and show none of the alteration to clays and iron oxides that is always found in terrestrial rocks. In addition, the lunar basalts contain more titanium (about 10 percent) than terrestrial basalts (about 2 percent). The lunar basalts also contain metallic iron as well as some new minerals that have never been found in earth rocks.

5 / The solid bedrock at the Apollo 11 landing site is covered by a layer of "lunar soil" a few meters thick composed of broken and melted rock fragments. Nearly all are basalt fragments from the bedrock underneath. This layer has been produced gradually by the continuous bombardment of meteorites for billions of years on the lunar surface.

6 / There are a few unusual white rock fragments in the lunar soil that are clearly different from the more common pieces of basalt. Such rare fragments may have been blasted out of the distant lunar highlands by meteorite impacts. If this is so, then the light-colored lunar highlands and the

dark lunar "seas" must be composed of different kinds of rock.

7 / Despite a major effort to find evidence of lunar life past or present no fossils, germs, or biogenic chemicals were found in the samples examined. The moon is "dead" and has apparently never supported life. Both biologists and romantics were disappointed, but these results were still important. We know now that life does not form everywhere and that it cannot form in any environment like the moon's.

As each new bit of information was extracted from the moon rocks, long years of debate were ended. Many old and staunchly defended theories toppled and new ones arose to take their place. Some of the new theories lasted only until November, 1969, when Apollo 12 returned with rocks from the Ocean of Storms.

The Apollo 12 rocks were also basalt lavas. Because two lunar "seas" had turned out to be made of this material, it now seemed likely that all the dark areas of the moon were basalt. But the Apollo 12 rocks from the Ocean of Storms were not the same as the ones returned by Apollo 11 from the Sea of Tranquillity. For one thing, they were about 3.3 billion years old, almost half a billion years younger than the Apollo 11 samples. Their chemistry was also different; they contained less titanium (about 5 percent) than the Apollo 11 basalts (10 percent), but still more than terrestrial basalts (2 percent).

Lunar theories had to be modified again to explain these differences. The idea of a hot, partly molten moon had to be expanded to include at least two distinct periods of melting, or perhaps instead a long period of heating followed by a flooding of lava that lasted half a billion years. The chemical differences between the Apollo 11 and 12 lavas indicated that the composition of the lunar interior differed from place to place. The two batches of lunar lavas had come, so to speak, out of different melting pots. The post-Apollo 12 moon had suddenly become more complicated than it had appeared after the first Apollo mission. It would continue to do so.

As the Apollo missions continued, the scientific results came in such a flood that it was difficult to keep up with them. The early launches had established that the Apollo system worked, so that missions to come could be devoted to more ambitious landings and a large science payload. The flat lunar "seas" were abandoned for more rugged, scientifically interesting sites. Improved television cameras carried armchair moonwalkers back on earth to the deep trench of Hadley Rille and to the steep slopes surrounding the Littrow Valley. The limited moonwalks of Apollo 11 and 12 were succeeded by long drives in a Lunar Rover. Samples came back to earth in batches of hundreds of pounds. More large rocks were returned, and long core tubes were driven into the lunar soil to sample its layers and bring them back to earth undisturbed.

As each pair of astronauts worked on the moon, a third, orbiting a hundred miles above them in the Apollo Command Module,* pointed instruments that photographed and analyzed wide bands of the lunar surface. And each time astronauts blasted off the moon, they left behind instruments that even today continue to send back to earth even more recent data on the vibrations, heat, and magnetic field of the moon.

Small lunar features, invisible through telescopes, became the stations for collecting moon rocks, and scientists created a new lunar geography with informal and sometimes

* The Apollo spacecraft consists of two parts, a Lunar Module (LM or LEM) in which two astronauts land on the moon, and a Command Module (CM) in which a third astronaut remains in orbit around the moon. At the end of the surface exploration, the LM blasts off the moon and joins the CM in lunar orbit. The astronauts transfer themselves and their samples from the LM to the CM, the LM is then discarded, and all three astronauts return to earth in the CM.

whimsical names: The Snowman, Cone Crater, Silver Spur, Smoky Mountain, The Slide, and Shorty Crater. Some of the more distinctive samples also acquired names, and scientists and reporters each presented their own versions of the silent information contained in Big Bertha, the Genesis Rock, House Rock, and the Orange Soil.

In the few hectic years since the landing of Apollo 11, we have learned more about the moon than in all the centuries preceding it. Sample collecting ceased in December, 1972 with Apollo 17, but study of the returned rocks still goes on, and new data from them are combined with the reports still coming back from the instruments on the moon. It took us a little less than ten years to build Apollo and get to the moon. It will take much longer before we understand all that we found there.

There is a great deal more to be learned about the moon in the future, but even now there is much that is clear. We now think that the moon formed at the same time as the earth and the rest of the solar system, about 4½ billion years ago. Chemically, the moon is similar to the earth, but there are also many chemical differences that make it hard to argue that the moon and the earth were ever part of the same body. Like the earth, the moon is a complex planet with a long history of change; but both bodies have evolved in different ways. The moon's early history is a time of great primordial heating, of catastrophic bombardment by huge asteroids, of great outpourings of molten lava, and finally, of three billion years of quiet. If the earth ever went through similar catastrophes when it was young, the records have been destroyed by the later episodes of volcanic eruption, mountain-building, and erosion that formed the rocks we now see around us.

As a by-product of exploring the moon, we have also been learning how to explore the rest of the solar system. The Apollo Program showed that the scientific tools originally developed to understand the earth could be successfully applied to determine the nature and history of the moon, and the techniques used to study the moon can now be applied to other planets. Even today, scientists have a better understanding of their new photographs of the surfaces of Mercury and Mars because of what they learned from the moon. And it will not be long before the moon, once the last word in inaccessibility, will serve as a routine testing ground for instruments intended for longer journeys.

CHAPTER

THE LONG WATCH

The moon was the timepiece that early man used to evolve from a wandering hunter to a settled farmer. The regular changes in the shape of the moon provided an interval of time intermediate between the day and the year. Using the waxing and waning cycles of the moon, man divided the 365-day year into 12 convenient periods; the shape of the moon during each month told him the time to within a day or so.

With the development of a lunar calendar, man became able to record the past and to plan for the future. He learned when to plant and harvest grain, thus developing a reliable food supply that freed him from his dependence on hunting skills and the abundance or scarcity of game. Permanent villages became possible, and civilization began.

Man is a practical animal, and he put the moon to good use without worrying about its nature. Planting by the moon became less common as written calendars and almanacs appeared, but the practical agricultural applications associated with the moon still persist in folklore and tradition. The waxing phases of the moon are associated with life and growth, and the waning phases with death and decay. In some countries, mothers still expose babies to the light of a waxing moon to give them strength, and trees are cut down under a waning moon when their vigor is supposedly reduced and the cutting is easier. The period between waning and waxing moons, called the "dark of the moon," is said to be a time of great potency for both good and evil deeds.

But man is also a curious animal, and even while he used the moon he tried to explain it. Nearly every culture, whether primitive or sophisticated, has developed explanations for various lunar phenomena, in particular the moon's light, its phases, the light and dark markings on its face, and the eclipses that occasionally occur.

Often the moon has been conceived of as a deity, or at least an object placed in the sky by one. In New Guinea, natives thought that the moon was a shining ball stolen from a child by the sun god who needed it to light up the night while he slept. The Babylonians saw the moon, which they named *Sinn*, as a male god, whom they credited with bestowing on man civilization and all forms of learning. A tribe in the Himalayas personified the moon as an amorous man whose unresponsive mother-in-law threw ashes in his face, producing the dark spots. Greenland Eskimos warned young girls that the moon was a god who would make them pregnant if they were unwise enough to sleep in the moonlight.

In many civilizations the moon was viewed as a female deity, possibly because it was natural to think of it as a sister or wife of the male sun. A stronger connection may have been the early discovery that the moon's 29½-day month exactly matched the menstrual cycle of women. The Greeks called the moon *Selene*, which is the root of our word *selenology*, the study of the moon. To the Romans she was *Luna*, from which derive such words as *lunar* and *lunacy*. Other mythological names for the moon, *Cynthia*, *Diana*, and *Artemis*, among them, provided variety for romantic poets through the centuries.

The light and dark pattern on the moon, commonly identified as "The Man in the Moon," has many mythological explanations. In European tradition, the "man" was banished from earth by God for a crime such as stealing cabbages on Christmas or collecting firewood on the Sabbath. He was also identified as Cain or Judas Iscariot.

Because of the curvature of the earth, people living near the equator see the moon as if it has been slightly rotated. To Australians, The Man in the Moon is completely upside down. The dark spots then form a different pattern, suggesting to Orientals a banyan tree or a hare. (The latter interpretation may be the source for modern ideas about the Easter Bunny and the luckiness of rabbits' feet.)

The moon has commonly been associated with weather, especially rain, and many different cultures recognized in the moon's dark spots the figures of people carrying water buckets which they emptied on the earth to produce rain. A Norwegian version of this belief still persists in the nursery rhyme about Jack and Jill and their pail of water.

The repeated cycle of the moon as it grew from a tiny crescent to a full moon, shrank again to a tiny crescent, vanished, and reappeared, was interpreted as the actions of a deity who turned her face toward and away from the earth, or who was born, died, and was born again.

Eclipses of the sun and moon occur when either the earth or the moon enters the shadow cast by the other. Eclipses of the moon, which take place when the moon enters the earth's shadow on the opposite side from the sun, are not spectacular and may pass unnoticed. They occur at night, and the total phase of the eclipse is over in about an hour. Furthermore, because sunlight is scattered through the earth's atmosphere onto the moon, the moon never disappears entirely, although its darkened part may take on a coppery or bluish color. This brief change in the moon, even if noticed, might be mistaken for the effects of a passing cloud. Although some myths speak of demons who occasionally swallow the moon, there is little indication that primitive people needed an explanation for such a subtle phenomenon.

By contrast, total eclipses of the sun can be sudden, obvious, and frightening. In this case, a small area of the earth enters the shadow of the moon, and the sun may be completely obliterated from the sight of anyone within the shadow, causing darkness in the middle of the day.*

If the sun is only partially covered by the moon, the reduction in light is barely detectable by the human eye. Even when as much as 98 percent of the sun is covered, the effect is a gradual twilight, and human activities go on much as usual. But for people in the path of the total eclipse, the transition from 98 percent to 100 percent coverage can be unnerving.

As the last light disappears, birds roost, animals sleep, and the temperature plummets. The landscape becomes covered with strange, rippling bands of light and shadow, and observers in high places see the shadow of the moon as a dark spot about a hundred miles wide, rushing toward them at a thousand miles an hour. Then there is sudden darkness, the temperature drops, and where the sun had been there appears a ghostly phosphorescent halo as the sun's normally invisible upper atmosphere shines out from

* The diameter of the sun is 1,384,000 kilometers (865,000 miles) or about 400 times the diameter of the moon, but the sun is also about 400 times further away than the moon. Because of this astronomical coincidence, the sun and moon appear to be the same size in the sky, each covering an angle of about half a degree. Therefore, the moon covers the disk of the sun more or less exactly.

behind the moon. Ancient peoples were terrified, battles suddenly ceased, kings and emperors abased themselves, and inattentive court astronomers lost their heads for not making accurate predictions. Both Orientals and American Indians believed that a demon, dragon, frog, or giant bird was eating up the sun and tried to frighten it away with drums, yells, dances, bells, or firecrackers. And everyone rejoiced when the sun appeared again, safe and sound.

Some of modern man's activities during total eclipses accidentally resemble those of his ancestors. On March 7, 1970, I stood on Chincoteague Island, Virginia, in the chilly darkness of a total eclipse, watching as the nearby NASA research station on Wallops Island launched some twenty small sounding rockets into the sky during the two minutes of total darkness. The rockets were carrying cameras and other instruments into the upper atmosphere, but the thundering noise of their motors and the fiery streaks that they made across the dark sky produced a spectacle that any ancient Chinese astronomer or Indian medicine man would have appreciated.

The question of how the moon affects the earth and its inhabitants has generated much myth, folklore, and serious scientific study. Tides and eclipses are two unquestioned effects of the moon on the earth. Aspects of the lunar calendar still survive in many ways, notably in fixing the date we celebrate Easter: the first Sunday after the first full moon following the Spring Equinox on March 21. There are definite but poorly understood relations between the phases of the moon and the mating activity of some marine animals. Sea urchins in the Mediterranean and the Red Sea spawn more actively at full moon. The palolo worms of the South Pacific rise to the surface to mate during the full moons of October and November, only to be caught by watchful natives waiting in canoes. More familiar to Americans is the grunion, a small marine fish that, for one night every year, swarms onto the beaches of Southern California in huge numbers to breed, providing Californians with an annual festival that is firmly fixed in the lunar cycle.

The possible effects of the moon on human beings are a major concern of legend and superstition. Folklore abounds with tales of werewolves and other moon-generated monsters. For centuries it was generally believed that the moon, especially the full moon, could cause madness, and so we have words like *moonstruck*, *mooning*, and *lunatic*. Belief in such myths has gradually declined as our knowledge about the mind has grown more sophisticated; any connections that exist between the moon and human behavior must be more subtle.

The coincidence between the 29½-day lunar month and the menstrual cycle of women is a well-established biological fact, but one that has never been fully explained. Research during the last 20 years has established that other biological functions in both men and women are also cyclical in nature. Human beings exhibit a cyclical regularity in such activities as sleeping and waking, mental alertness, in the bodily production of certain chemicals, and in the degree of response to drugs. About a hundred bodily functions have been found to have cycles that approximate the 24-hour length of the day (circadian rhythms). A few other body activities have 15-day and 30-day periods that may be related to the lunar cycles. This kind of research, called "chronobiology," is just beginning, and it is too early to tell whether the idea of lunar influences, like many other folk beliefs, actually rests on some undiscovered scientific facts.*

The early history of man reflects his long preoccupation with the moon and his efforts to keep track of it. More than 30,000 years ago men recorded the phases of the moon in

* Two good sources for more information about chronobiology are *The Living Clocks* by Ritchie Ward (New York: Knopf, 1971) and *Body Time* by Gay Gaer Luce (New York: Bantam, 1973).

bone carvings. By 2000 B.C. the Babylonians had established observatories, recorded the motions of the moon and the five visible planets, and learned to predict eclipses accurately. Further north, at about the same time, the inhabitants of Britain built the great megalith at Stonehenge for the same purpose.

Between about 400 B.C. and 150 B.C. the Greek philosophers and geometers developed the tools needed to study the solar system and also produced some surprisingly modern theories about its nature. Anaxagoras and Aristotle taught that the moon was a solid object illuminated by the sun, which was an advance over earlier theories that it was a fiery substance or a thick cloud. Anaximander and Pythagoras established the idea that the earth was a spherical heavenly body and that it was in fact the center around which the rest of the visible universe revolved. By contrast, Aristarchus argued that the earth and other planets circled the sun, and Nicetas (or Hicetas) of Syracuse suggested that the earth itself rotates, producing the illusion that the sun and stars move around it.

There were limits to Greek enlightenment however. Anaxagoras was supposedly condemned to death (although not executed) for proposing the fiery nature of the sun and for arguing that the earth and moon might be made of the same kind of material. And the heliocentric theory of Aristarchus and Nicetas fell out of favor and lay forgotten for sixteen centuries until Nicolas Copernicus revived it again.

The unique contribution of the Greeks was their application of geometry to obtain the first basic measurements of the universe. Aristarchus attempted to determine the relative distances of the moon and sun from the earth. Eratosthenes calculated the size of the earth to better than 10 percent of the correct value, thus providing a fundamental figure from which the moon's size and distance could be accurately calculated. Hipparchus calculated the distance to the moon to within 10 percent accuracy. Considering the large uncertainties in the crude basic measurements, such accuracy was a major achievement. Hipparchus also noted the positions of about 1,000 stars, thus founding the science of observational astronomy. Another fifteen hundred years would elapse before man climbed back up from the Dark Ages to the same level of understanding about his universe.

The studies of the Greeks culminated in the Ptolemaic System, developed by a brilliant theorist, Claudius Ptolemy (or Ptolemaeus), about 150 A.D. In this system the earth was fixed at the center of the universe and the sun, moon, and planets revolved around it in complicated orbits made up of multiple circles. For the Greeks the circle represented the perfect, divine geometric shape.

Much is made of the "overthrow" of the Ptolemaic System in the Renaissance, but it is simplistic to regard this revolution as the triumph of Copernicus's "good" theory over Ptolemy's "bad" one. The Ptolemaic System was a cumbersome and arbitrary construction with no basis in physical laws. But it did combine centuries of observation of planetary motions with a reasonable and well-developed theory. For 1,500 years it met every test of observation that could be mustered, and it predicted the motions of the planets with all the accuracy necessary during that period. It proved quite adequate for a Europe where "science" was dominated by dogma, Aristotle, and astrology; it also worked perfectly well in the Islamic countries where the sciences, including astronomy, continued to flourish.

Furthermore, the chief advantage of the Copernicus–Kepler model was artistic rather than observational; in this theoretical model the orbits of the planets formed simple ellipses rather than complex combinations of circles. The predicted planetary movements did not differ a great deal from the Ptolemaic System. The critical difference in the observed position of Mars was about one-eighth of a degree, which is roughly equivalent to one-quarter the width of the full moon. This difference was too small to have been detected by Ptolemy or, for that matter, by Copernicus himself.*

Although Copernicus suggested as early as 1506 that the earth revolved around the sun, the decisive evidence was only produced much later, in the seventeenth century when the careful observations of Tycho Brahe were combined with the mathematical genius of Johann Kepler.

Almost immediately after Kepler's work on the elliptical orbits of the planets appeared in 1605 technology spurred a major upheaval in the study of the solar system. In Holland, lens-makers had discovered how to combine a series of glass lenses in a tube to produce magnified images of distant objects; a few years later, after details of its construction reached Italy, the telescope came to the attention of a Professor of Mathematics at the University of Padua named Galileo Galilei. He quickly reproduced the invention and, after trying it on buildings and distant ships, swung it around and looked into the sky.

Galileo's "spyglass" seems almost insignificant by modern standards. The largest one he constructed magnified only about 30 times, providing a view of the moon not much better than what a modern viewer might see through a good pair of binoculars. But Galileo's work marks the beginning of planetary astronomy just as Hipparchus's observations inaugurated the scientific study of the stars. The solar system expanded and took on unsuspected details; the planets became more than wandering lights in the sky. Galileo's telescope revealed the phases of Venus and the four large moons that circle Jupiter like a miniature solar system. And it also brought the moon closer to man than it had ever been before.

Through the telescope the moon became, once and for all,

* Two excellent sources for further reading in the history of astronomy are *And There Was Light* by Rudolf Thiel (New York: Mentor, 1960) and *The Sleepwalkers* by Arthur Koestler (New York: Grosset and Dunlap, 1963).

another world apart from the earth. The Greeks had believed this, but Galileo went on to show that the moon had its own unique geography and surface features. The moon was a rough world, with rugged mountains and hundreds of depressions which Galileo called "small spots," thus giving us the first description of lunar craters. He observed that the dark patches that form the eyes and mouth of The Man in the Moon were more level and smoother than the light regions. Suggesting that the dark regions might be similar to earth's oceans, he called them *maria* (plural, pronounced MAH-ree-ah; singular *mare*, pronounced MAH-ray)—Latin for "seas," a term that scientists and astronauts still use. The rough, bright regions now called "highlands" or "uplands" he named *terrae*, or "lands."

But Galileo was an aggressive and articulate publicist as well as a brilliant experimental scientist. He was not the only scientist studying the moon at this time—Thomas Harriott in England had made lunar observations and maps just before Galileo but never published them. Galileo, however, published his observations quickly, using language that was unusually clear and understandable. In *The Starry Messenger*, which appeared in 1610, he presented his painstaking observations of the solar system in an exciting and readable style and articulated for the first time what became the modern view of the nature of the moon:

> For greater clarity I distinguish two parts of this surface, a lighter and a darker; the lighter part seems to surround and to pervade the whole hemisphere, while the darker part discolors the moon's surface like a kind of cloud, and makes it appear covered with spots. . . . From observations of these spots repeated many times I have been led to the opinion and conviction that the surface of the moon is not smooth, uniform, and precisely spherical as a great number of philosophers believe it (and the other heavenly bodies) to be, but is uneven, rough, and full of cavities and prominences,

being not unlike the face of the earth, relieved by chains of mountains and deep valleys.*

In the half-century after publication of Galileo's findings numerous observers made the first complete lunar maps. In 1645 Langrenus produced a map showing about 300 lunar features, 250 of which were prominent craters. An astronomer at the court of King Philip II of Spain, Langrenus named the features of the lunar surface after Spanish kings and nobles. These names have not survived, with one notable exception—the crater he named after himself.

Hevelius's map, published in 1647, featured lunar mountains with designations after mountain ranges on earth, among them, the Alps, Apennines, and Carpathians. However, it was Riccioli, in 1651, who established most of the conventions that we use today for naming surface features on the moon. The dark maria were given Latin names indicating qualities or characteristics. Some of the maria that became important to the Apollo Program are: *Mare Imbrium* (the Sea of Rains), *Mare Tranquillitatis* (the Sea of Tranquillity), *Mare Serenitatis* (the Sea of Serenity), *Mare Orientale* (the Eastern Sea), and *Oceanus Procellarum* (the Ocean of Storms).

Riccioli also began the current practice of naming lunar features, particularly the craters, after philosophers, scientists, and historical figures (Tycho, Copernicus, Kepler, Plato, Aristarchus, and Julius Caesar). Today this tradition has been formalized by the International Astronomical Union (IAU), and craters are still named after deceased scientists and historical figures, especially those active in astronomy and the study of the moon (Herschel, Yerkes, Abbott) or in the exploration of space (Tsiolkovsky, Goddard, Gagarin, Gast).†

* As quoted in Stillman Drake, *Discoveries and Opinions of Galileo* (New York: Doubleday Anchor, 1957), p. 31.

† Sir William Herschel (1738–1822), a British astronomer, de-

While seventeenth-century scientists mapped the moon, writers were busy constructing a variety of fanciful moons and inventing miraculous ways of reaching them. Science and science fiction have always gone hand in hand where the moon was concerned.

Johann Kepler's *Somnium* (or *Dream*) was published in 1634, shortly after the great scientist's death. As might be expected, Kepler's dream-voyage to the moon incorporates a number of noteworthy scientific insights. He describes the earth fixed permanently in the lunar sky, the alternation of scorching two-week "days" and equally long freezing "nights" in addition to some effects that foreshadow Newton's Law of Universal Gravitation.

This speculative literature evolved as our knowledge of the moon expanded. By 1700, observations had shown that life on the moon was unlikely; there was no significant atmosphere, no liquid water, and a fatal cycle of too-hot "days" alternating with too-cold "nights." Newton's Laws of Motion had put firm limits on the physical forces that one could use to get to the moon, and it seemed likely that most of the space between the earth and the moon was an airless vacuum in which no traveler could live.

The most exciting and widely read lunar literature of the 19th century was a thorough hoax, a piece of pure fantasy wrapped in a thin layer of scientific fact, "The Great Moon Hoax" of 1835.

---

veloped reflecting telescopes and discovered the planet Uranus. Charles T. Yerkes (1837–1905) was an American businessman who helped build the famous Yerkes Observatory in Williams Bay, Wisconsin. Charles G. Abbott (1872–1973) was a long-lived solar astronomer and former Secretary of the Smithsonian Institution. Konstantin E. Tsiolkovsky (1857–1935) and Robert H. Goddard (1882–1945) were, respectively, Russian and American pioneers in the development of rockets for space travel. Russian cosmonaut Yuri A. Gagarin (1934–1968) became the first man to orbit the earth in 1961; he was later killed in a plane crash. Paul W. Gast (1930–1973), a prominent scientist, supervised geochemical studies of lunar samples for NASA until his death from cancer in 1973.

In August, 1835, The New York *Sun* began a series of articles, allegedly reprinted from a nonexistent journal called the "Edinburgh Journal of Science," which stated that Sir John Herschel, a distinguished British astronomer and the son of Sir William Herschel, was making astronomical observations in South Africa. That much was true; the rest of the whole series was the creation of *Sun* writer Richard Adams Locke, who described in detail an imaginary telescope that weighed 77 tons, magnified 42,000 times, and brought the moon to an apparent distance of a few hundred yards. Through the telescope, Locke wrote, the observer could see strange beasts and huge gems amid lunar beaches, grasslands, and waterfalls. He soberly described miniature bison, horned bears, and tailless beavers, finally identifying intelligent lunar inhabitants who "averaged four feet in height and were covered except on the face with short and glossy copper-colored hair and had wings composed of a thin membrane."

This fantasy was widely accepted as fact by a population that was favorably disposed toward the idea of life on the moon. The Moon Hoax produced an explosion of public interest and excitement that would not be matched again until the Apollo missions actually flew. The *Sun*'s circulation shot up above that of every other paper in the country and, finally, in the world; on August 28 it exceeded even the sales of the *Times* of London with a fantastic (for those days) 20,000 copies. A pamphlet version of the article sold another 60,000 copies, and even though the hoax was exposed after a few weeks, many people continued to believe firmly in the existence of winged moonmen.

For all its impact on the public the Moon Hoax was mere fancy compared to the more scientifically reasonable lunar voyages described by nineteenth-century writers like Jules Verne and H. G. Wells. Verne's classic *From the Earth to the Moon* and its sequel, *Round the Moon*, published in English translations in 1873, describe the construction of a huge gun to shoot a projectile containing three men to the moon. The propulsion system would not have worked (the acceleration would have been about 30,000 times the force of gravity, and neither projectile nor astronauts would have survived), but the books are otherwise surprisingly modern. Verne stated the correct escape velocity (7½ miles per second) and also projected the use of test animals and the effects of weightlessness. His voyage is astonishingly similar to the Apollo 8 mission more than a century later: both had a crew of three astronauts, a launch from Florida, a trip around the moon, and a safe splashdown in the Pacific Ocean.

H. G. Wells's *The First Men in the Moon* envisioned the use of a new technology (antigravity material) to carry two explorers to the moon, where they find both plants and hostile life. The novel was published in 1901 as the world entered the twentieth century; the actual landing of two men on the moon was less than a single human lifetime away.

CHAPTER

THE ORIGIN OF THE MOON: "ALL EXPLANATIONS ARE IMPROBABLE"

By the beginning of the twentieth century the physical features of the moon had become well known, although questions about its origin and history were still unsettled. Its size, weight, and density had all been accurately determined. Its motions had been precisely observed, and its location could be predicted for millions of years to come.

The basic statistics of the moon have been well established for the last 75 years. The moon swings around the earth in a nearly circular orbit that is about 382,000 kilometers (238,000 miles) away. This is not a great distance; an active executive might travel that far in less than two years. The moon is a sphere whose diameter is 3,500 kilometers (2,180 miles)—about equal to the distance between New York and El Paso, Texas, or between St. Louis and San Francisco. The surface area of the moon is about 38 million square kilometers (15 million square miles)—nearly that of North and South America combined.

Although the diameter of the moon is about one-quarter

that of the earth, the moon weighs only about one-eightieth as much as the earth. The force of gravity at the moon's surface is only one-sixth that of the earth. A fully suited astronaut weighing about 350 pounds on the earth weighs only about 60 pounds on the moon.

Out of these basic statistics emerges a fundamental difference between the earth and the moon. The moon's density (mass divided by volume) is 3.35 grams per cubic centimeter (water weighs 1.0 gram per cubic centimeter), whereas the density of the earth is 5.5. The fact that the earth is 60 percent denser than the moon suggests that there is some basic difference in their chemical composition—a difference hard to explain in two bodies that are so close together in space.

There is another interesting aspect to the different densities of the earth and moon. The earth's outer layer of silicate rock is about 3,200 kilometers (2,000 miles) thick and has an average density of about 3.3 which, as it happens, nearly equals that of the moon. This coincidence in densities is a basic fact of lunar geology, one that all theories of lunar origin have had to struggle with. The earth also has a much denser central core of metallic iron, but apparently the moon is not heavy enough to have a large core like the earth's.

The moon has no atmosphere. When stars pass behind the moon they disappear sharply and suddenly with none of the gradual dimming that would be produced if their light was passing through a lunar atmosphere. More recent studies have shown that natural radio sources in the sky are cut off in the same sudden way as the moon moves in front of them. These movements show that at the lunar surface there is a more complete vacuum than can be produced in any terrestrial laboratory.

The absence of a lunar atmosphere is not surprising; the moon's gravity is too weak to hold an atmosphere like the earth's. If relatively light gases like oxygen, nitrogen and water vapor were ever present on the moon, their molecules must have escaped into space long ago.*

This lack of an atmosphere means that, unlike the earth, the surface of the moon has no protection from continuous bombardment by tiny meteorites and from scorching by lethal X-rays, gamma rays, and cosmic rays that emanate from the sun and the rest of the universe. Fortunately for us, this dangerous matter and energy is absorbed by our atmosphere before it reaches the surface of the earth.

The moon completes one orbit around the earth in 27.3 days (the sidereal month). However, the earth also moves along its orbit around the sun while the moon is swinging around the earth. As a result, the angle of illumination of the moon by the sun changes slightly, and a longer period passes before the moon returns to the same phase as seen from the earth. This latter period, the time between one full moon and the next, is 29.5 days (the synodic month), and it has long been the basis of the lunar calendar.

The moon is also "locked" in its orbit, and as it moves around the earth, it turns so slowly that it always keeps the same side facing toward the earth (*Figure A*). The moon thus rotates once on its axis in the same time that it makes one trip around the earth. To keep one face turned always to the earth, the moon must turn its back on the sun during half its orbit. (Anyone can demonstrate this effect by walking in a circle around some object like a lamp or a birdbath while keeping it continuously in sight.)

As a result of these motions, the 29½-day month is divided on the moon into a lunar "day" and a lunar "night," each about two weeks long. Because the moon has no insulating atmosphere, the "daytime" temperature in direct sunlight is about 134° C. (270° F.), well above the boiling

* The moon's gravity is strong enough to hold back heavier atoms like argon and radon, but there are not enough of these elements present to make any tangible atmosphere. See page 204.

point of water. During the lunar "night," the temperature drops suddenly to about −170° C. (−270° F.), much colder than the freezing point of carbon dioxide ("dry ice").

The moon's surface has been mapped in greater detail than many parts of the earth. Observers in the early twentieth century had available the excellent maps, made with small telescopes, of Beer and Mädler (1850), and Schmidt (1878). As techniques improved, these maps were followed by the more detailed efforts of such observers as Wilkins (1935), and Fauth (1964). Around 1960 the U.S. Air Force Aeronautical Chart and Information Center began an ambitious and ultimately successful program to map the whole visible side of the moon on a scale of 1 to 1,000,000.

These maps, even the most modern ones, were drawn almost entirely from direct visual observations. The first photograph of the moon was taken about 1840, almost immediately after photography itself had been invented. However, photography did not become a major part of lunar mapping until the advent of the Space Age, despite the fact that the Lick Observatory in California produced

---

Figure A / The Light and Dark of the Moon. The moon's phases (A) and eclipses (B) result from its illumination by sunlight at various points in its orbit around the earth. (The drawings are not to scale, and the moon, earth, and sun are actually much farther from each other than shown here.)

"New Moon" (A) occurs when the moon is between the earth and the sun, so that the earth-facing side is completely dark. "Full moon" occurs on the opposite side of the moon's orbit when the earth-facing side is fully lit. (The dark spots represent the lunar maria that form "The Man in the Moon.") As the moon moves around the earth, it turns on its axis to keep the same face toward the earth. The moon thus rotates once on its axis while revolving once around the earth.

Eclipses (B) occur only when the moon, earth, and sun are in a straight line. Because the orbit of the moon is slightly inclined to the earth's orbit around the sun, such an exact line-up occurs only rarely. An eclipse of the moon (upper) occurs when the moon passes into the shadow of the earth. An eclipse of the sun (lower) is produced when the moon's shadow passes across part of the earth's surface.

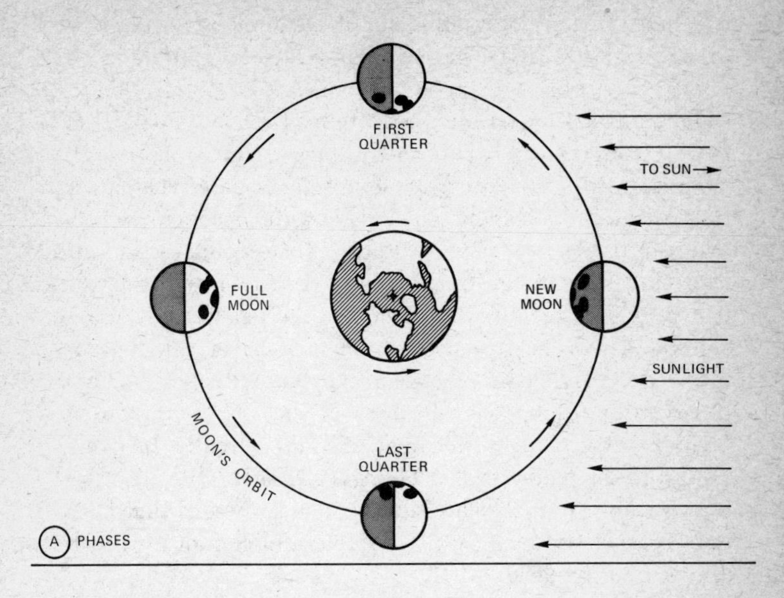
FIRST
QUARTER
TO SUN
FULL
MOON
NEW
MOON
SUNLIGHT
MOON'S ORBIT
LAST
QUARTER
A PHASES

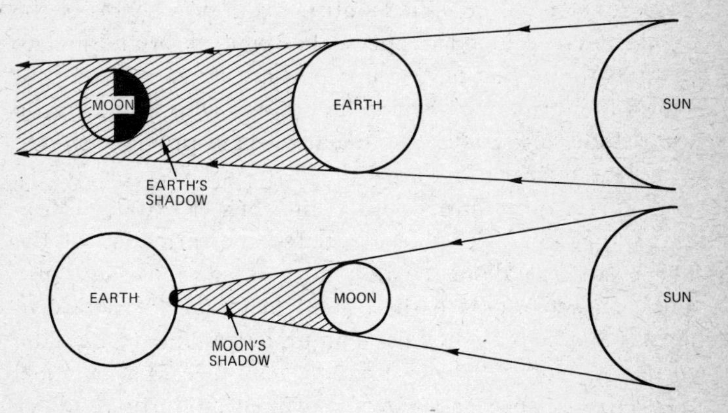
MOON
EARTH
SUN
EARTH'S
SHADOW
EARTH
MOON
SUN
MOON'S
SHADOW

B ECLIPSES

excellent lunar photographs about 1880 and several photographic atlases of the moon were available by the turn of the century.

Lunar mapping requires the ability to discriminate small details from the earth, and for this purpose the human eye is superior to the camera. A camera records only what it sees during the brief instant when the shutter is open. But the eye of a patient observer, watching for seconds or minutes at a time, can take advantage of fleeting instants of unusually clear "seeing" that occur as the earth's atmosphere shifts and wavers. Because a human observer can thus detect objects much smaller than can normally be seen in a photograph, direct visual observations of the moon remained important until cameras could actually be transported to the moon itself. (There is a disadvantage to visual observations, too: the descriptions and drawings that a visual observer produces are subjective and cannot always be independently checked.)

While scientists compiled statistics and drew maps of the moon, they continued to speculate about its origin, and to try to fit the formation of the moon into a developing concept of the formation of the solar system. During the last 30 years, studies in various scientific fields have provided significant data relating to the birth of the solar system itself. Geological studies of meteorites, tiny bits of extraterrestrial matter that fall to earth, have dated the formation of the solar system at about 4.6 billion years ago and have provided analyses of its earliest solid material. The study of nuclear reactions has led us to an understanding of how the sun burns and what its lifetime may be. Mathematical models have been constructed to explain the birth and later evolution of the planets.

After centuries of speculation and debate, a scientifically based theory of the origin of the solar system has finally emerged, called the "solar nebula" or "dust cloud" theory. In this theory , the earliest recognizable form of the solar system is a huge, disk-shaped mass of slowly whirling gas and dust (the solar nebula) that came into existence about 4.6 billion years ago. Gradually, the gas and dust spiraled inward, and finally the innermost part of the cloud became dense enough and hot enough to form the sun. Farther away from the center of the cloud, where it was cooler, the dust and small particles accumulated at regularly spaced intervals to form planets. Eventually the solar system assumed its present form: one sun, 9 planets, 34 moons, and uncountable small asteroids, meteorites, and comets. How long this process took is not clear. The evidence from meteorites indicates that it could not have been longer than a few million years; it might even have been less than a few thousand.

The "dust cloud" theory still has many unresolved problems. The tiny dust particles have practically no gravitational force to attract each other—so what mechanism could have caused the first few dust particles to stick together and build up larger bodies? Why are there two kinds of planets—small, dense ones (Mercury, Venus, Earth, Mars) close to the sun, and giant, gas-rich ones (Jupiter, Saturn, Uranus, and Neptune) farther out? Why is the outermost planet, Pluto, small and dense instead of large and gaseous like its neighbors? If the entire original dust cloud was whirling rapidly, how did it happen that the central sun, which contains 99.9 percent of the total mass of the solar system, ended up with only 2 percent of the angular momentum* while the planets somehow collected all the rest?

Furthermore, the earth-moon pair is an anomaly in the solar system. Our moon is very large as moons go, almost as

* The *angular momentum* of an object or group of objects rotating around an axis is defined by the product mvr, where m is the mass, v is the rotational velocity, and r is the distance from the axis of rotation. A fundamental law of physics is the Conservation of Angular Momentum, which requires that the total mvr remain constant even though the individual terms may change. Thus, as the original solar nebula contracted (r decreases), it began to spin faster (v increases

large as the planet Mercury, but it circles around a relatively small planet. This arrangement is unique in the solar system. The other earth-size planets have no moons at all, except for Mars, which has two small ones (Phobos and Deimos) only a few miles in diameter. Satellites as large as our moon are found only around the giant planets. Four of Jupiter's 12 satellites (the four first seen by Galileo) are close to the moon in size; Io and Europa are slightly smaller and Ganymede and Callisto are larger. Titan, the largest of Saturn's 9 moons, is 50 percent larger in diameter than our own moon, and massive enough to retain some kind of frozen atmosphere.

To an observer standing on Mars or Venus, the earth and moon would appear together as a unique "double planet," one blue, one white. The smaller moon would reflect the sunlight and appear white. The larger earth would appear blue because of the reflection of sunlight from the oceans and the scattering of sunlight in the atmosphere. (This scattering of sunlight is what makes the sky appear blue to us.)

The unusual character of the earth and moon has made it difficult to fit them into a single acceptable theory for the formation of the solar system. As scientists continued to observe and speculate, three separate theories, all equally reasonable, were proposed to explain the origin of the moon.

According to the "fission" or "escape" theory, the moon is a "daughter" of the earth. This theory suggests that a single large planet formed where the earth is now; and as it cooled, it began to spin so rapidly that it flattened into a disk-shape, then into a sausage-shape, and finally split into two

---

to balance the decrease in r). A more familiar application of this principle is seen in a spinning figure skater, who spins slowly with his arms extended and spins faster when he pulls them in toward his body.

parts. The larger part became the earth, and the smaller part was flung out into orbit to become the moon.

The "double planet" theory suggests that the earth and moon condensed into separate bodies as the solar system was forming. The moon is thus a "sister" of the earth; both bodies formed close together, and they have remained close together ever since.

The "capture" theory views the moon almost as a "girl friend" of the earth. The moon formed separately in another part of the solar system, possibly in the asteroid belt, or perhaps outside the solar system entirely. Some time after it was formed, it passed close enough to the earth to be captured by the earth's gravity and held in orbit around it. This theory doesn't explain how the moon formed; it just explains how it got to where it is now.

The fission theory was proposed about 1880 by G. H. Darwin (1845–1912), son of the famous British biologist Charles Darwin. The younger Darwin was an authority on the nature and effects of tides produced on the earth by the sun and moon, and his theory evolved from a demonstrated fact: because of the action of tides on the earth, the moon is now moving slowly away from the earth, and it must therefore have been much closer to the earth in the past.

Nearly everyone is familiar with the rise and fall of the tides along the seashore, and most people know that tides are caused by the attraction of the moon and, to a lesser extent, of the sun. The largest tides are in the earth's oceans; a rise and fall of about 75 centimeters (2½ feet) is produced in the open sea, and even higher tides occur along some coastlines. The moon also produces tides in the solid earth as well, but rocks are much more rigid than water, and the movement produced, about 11½ centimeters (4½ inches) needs more careful measurements to be detected.

A less obvious effect of tides is that they cause the earth to slow down. The friction of the tides along the coastlines and the floors of shallow seas dissipates about two billion horsepower of energy and acts as a brake on the rotation of the earth. The effect is slight but real: each day is about five hundred-millionths of a second longer than the one before. This is an insignificant amount, but like compound interest, it adds up as time passes. In only a century, this small change produces an error of about 33 seconds, which is large enough to be detected in the timing of ancient eclipses.

As the earth slows down, the moon must move farther away so that the total angular momentum of the earth–moon system remains constant. Modern calculations suggest that the moon is moving away from the earth at a rate of about 2 centimeters a year or about 2 meters a century.

Ancient rocks on the earth have recently provided evidence that the slowing down of the earth and the outward movement of the moon have been going on for a large part of the lifetime of the earth. In 1963 J. W. Wells (1907–) described growth structures in fossil corals collected from sedimentary rocks 400 million years old. He recognized a yearly growth cycle which he subdivided into about 400 daily cycles, suggesting that the year contained about 400 days and that each day was therefore about 21½ hours long instead of the present 24.

Projecting the observed motion of the moon back through time, Darwin reached a point when the moon was about 16,000 kilometers (10,000 miles) away from the earth and revolved around it in 5½ hours. (The earth also rotated on its axis in the same period, so that the day and month were both 5½ hours long.) Darwin went beyond this stage to suggest that the moon had actually broken away from the earth as a result of large tides raised by the sun on the rapidly rotating planet.

The fission theory, however, suffers from the absence of a convincing mathematical model for the actual separation process and from uncertainties about what the mechanical and physical properties of the original planet might have been. The observed recession of the moon from the earth does not necessarily prove that fission ever occurred; the motion can be explained by other theories of the moon's origin.*

Despite its drawbacks, the fission theory made it possible to predict what the moon's rocks might be like. The theory required that the separation of earth and moon had to take place at about the time the earth formed, 4.6 billion years ago, so that the moon must be over 4 billion years old and possibly 4.6 billion years old, or as old as meteorites themselves. Furthermore, the moon should have broken off from the outer part of the original earth. The moon should therefore be composed of silicate rocks like those that make up the outer part of the earth, and it would have little of the iron and other metals that had collected into the earth's core before the moon broke away.

The "double planet" theory evolved out of the more modern view that the solar system formed from a primordial whirling dust cloud, with the earth and moon now occupying the place where two planets condensed instead of one. Although this idea is consistent with the "dust cloud" theory of the origin of the solar system, there are still difficulties with it. Why should two planets have formed instead of one? Astronomical calculations indicate that a body that grows to the size of the moon will have a strong enough gravitational field to sweep up and collect all the smaller solid particles around it. Thus, once a small planet reaches a certain size, it prevents anything else from starting to grow in its neighborhood. Put another way, if a single body had grown to the size of the moon anywhere in the earth–moon

* The astronomer W. H. Pickering (1858–1938) proposed a later variation of the fission theory in which the moon was supposed to have come out of the Pacific Ocean basin. This theory, never widely accepted, has now been totally disproved, most recently by geological evidence from deep-sea drilling that shows that the ocean basins cannot be older than about 200 million years—much too young to have been involved in the formation of the moon.

region, then no other body would have been able to form anywhere near it.

Even if this mechanical problem had somehow been overcome, the apparent difference in composition between the earth and the moon is another stumbling block for the "double planet" theorists. Why should two bodies of different composition form at nearly the same place in the original dust cloud? A number of explanations have been offered, most of which involve differences in solidification temperatures or complex relations between chemical reactions and magnetic or gravitational forces. These explanations added so many complications to the basic theory that some scientists began to wonder if there might be an entirely different and much simpler explanation for the origin of the moon.

The composition of the moon was harder to predict if it had formed as a "double planet" with the earth. The moon might show chemical similarities to the earth. It might also resemble the small meteorites that had formed in the solar system at the same time. The moon might be a preserved and unchanged sample of some of the material that had formed the earth.

The third theory, that the moon has been captured by the earth after forming somewhere else, makes it possible to explain any differences in chemical composition between the moon and the earth. The original dust cloud probably did not have the same chemical composition everywhere, and a body that formed in the asteroid belt or near the orbit of Venus would probably be different from the earth.

If the moon had been captured, it is possible to estimate when the capture might have taken place. The moon's orbit and its gradual motion away from the earth are well known. If these motions are calculated back over the long span of geologic time, it appears that the moon's orbit and the earth–moon distance went through sudden and drastic changes about 1½ to 2 billion years ago. This sudden change, it is suggested, represents the time of capture of the moon. The implication of this theory is that the earth–moon system is not as old as the earth itself, and the moon was captured long after much of the earth's still-preserved rocks had been formed.

However the capture theory also entails difficult problems. For one thing, capture is a very unlikely event; a moon-size body passing close to the earth would be much more likely to smash into the earth or miss it completely. Furthermore, there are a number of mechanical problems. To slow down a body going by at several kilometers a second and bring it into orbit around a planet, strong forces must be applied and a great deal of energy must be removed from the passing body.* Finally, the forces involved in the capture would have produced catastrophic changes on both the earth and moon. But the earth's preserved rocks show no indication of abnormal widespread melting, earthquakes, floods, or volcanic eruptions about 2 billion years ago. In fact, geologic evidence indicates that there have been oceans and tides for at least 3½ billion years, which means that the moon has been around much longer than capture calculations suggest.

Even within the framework of the capture model, the chemical nature of the moon was impossible to predict; coming from anywhere the moon might literally be any-

* The space program provides a number of modern examples of the capture process. The Apollo spacecraft required a major rocket burn to slow down enough to move from an earth–moon trajectory into an orbit around the moon. Without this burn, the velocity would be so high that the spacecraft would swing around the moon and head back to earth or out into space. The Pioneer 10 and 11 unmanned probes did not try to slow down on their recent flybys past Jupiter; as a result, they swung around Jupiter, picked up *more* energy from Jupiter's gravitational field, and headed away from Jupiter much faster than they had approached it.

thing. Nevertheless, this theory made the moon a more interesting object. It might be an asteroid, a collection of matter from the neighborhood of Venus or Saturn, or even a fragment from beyond the solar system entirely. If the moon turned out to be a primordial object except for one sudden and violent episode of heating and lava production about 2 billion years ago, then the capture theory would be the best explanation.

But until actual moon rocks could be analyzed and their ages determined, all the theories to explain the moon's origin depended on uncertain mechanisms, unprovable assumptions, intricate mathematical calculations, and the risky projection back over billions of years of data from only a few centuries of observation. None of the three theories was particularly convincing, and all the explanations seemed uncertain and unlikely. H. C. Urey (1893–   ), the Nobel-Prize-winning chemist who became one of the most active speculators about the origin of the moon, summed up the uncertainty by saying, "All explanations for the origin of the moon are improbable"—which is another way of saying that he didn't like any of them very much.

It is natural that the uncertainty about the origin of the moon was reflected in debate and disagreement over the origin of its surface features as well. Everything observed on the moon provoked several contradictory explanations.

Arguments over the origin of the numerous lunar craters began almost immediately after they were first observed by Galileo in 1610. One explanation held that the craters were formed by explosive volcanic eruptions. Another held that the craters were produced by the bombardment of the moon by meteorites and asteroids.

The basic issue was stated as early as 1665 by Robert Hooke (1635–1703), an English naturalist who not only observed the moon but, in a surprisingly modern approach, made experiments trying to duplicate the lunar craters on a small scale. Writing in his *Micrographia* (1665), Hooke demonstrated the possibilities of heat and volcanic action by boiling a mixture of wet "alabaster" (equivalent to modern plaster of Paris) and observing the results:

> . . . the most notable, representation was, what I observ'd in a pot of boyling Alabaster, for there that powder being by the eruption of vapours reduc'd to a kind of fluid consistence, if, whil'st it boyls, it be gently remov'd besides the fire, the Alabaster presently ceasing to boyl, the whole surface, especially that where some of the last Bubbles have risen, will appear all over covered with small pits, exactly shap'd like these of the Moon, and by holding a lighted Candle in a large dark Room, in divers positions to this surface, you may exactly represent all the *Phaenomena* of these pits in the Moon, according as they are more or less inlightned by the Sun.*

In other experiments Hooke demonstrated that impacts could also form craters. He used

> . . . a very soft and well-temper'd mixture of Tobacco-pipe clay and Water, into which, if I let fall any heavy body, as a Bullet, it would throw up the mixture round the place, which for a while would make a representation, not unlike these of the Moon; but considering the state and condition of the Moon, there seems not any probability to imagine, that it should proceed from any cause *analogus* to this; for it would be difficult to imagine whence those bodies should come; and next, how the substance of the Moon should be so soft.†

We cannot fault Hooke for his uncertainty about where impacting bodies could come from to hit the moon. In Hooke's time, no one knew that comets moved around the

* Robert Hooke, *Micrographia* (New York: Dover, 1961), p. 243.
† *Ibid.*, p. 243.

sun or that asteroids even existed. Almost a century and a half would pass after the *Micrographia* was published before a massive and well-documented meteorite shower striking a small town in France finally convinced scientists that stones from outer space did strike the earth and could have hit the moon as well.

The debate between the impact and the volcanic origins of lunar craters has continued for over 300 years, and it is by no means settled even now. Even before the Apollo landings, most scientists were willing to admit that there were small craters of both kinds on the moon. The "volcanists" had to concede that the surface of the airless moon must be struck by meteorites that would form small craters. Conversely, most "impacters" admitted that there were some small lunar craters that looked volcanic. These craters are less than a few kilometers in diameter, or about the size of terrestrial volcanoes. They lack any uplifted rim and have no ejected material around them, and they are often arranged in lines or regular patterns that randomly impacting meteorites would be unlikely to produce.

The real debate between "impacters" and "volcanists" has raged around the question of whether the large lunar craters and even larger mare basins were formed by the impacts of asteroids miles in diameter or by volcanic eruptions on a scale never seen on earth. This argument has been pretty much of a stand-off; one reason is that the geology of the earth provides evidence to support both theories.

Volcanic action is widespread on the earth, and geologists can point to a variety of volcanic features that resemble lunar craters. There are numerous volcanic mountains like Vesuvius (Italy), Fujiyama (Japan), Paricutin (Mexico), Kilauea (Hawaii), and hundreds of others in Iceland, Siberia, and South America. These volcanoes appear similar to the central peaks observed in some large lunar craters. In addition, the earth also exhibits many volcanic collapse depressions (calderas), some of which are 10 to 30 kilometers in diameter and closely resemble many lunar craters. (Crater Lake, Oregon, about 10 kilometers in diameter, is a well-preserved caldera familiar to most American tourists.)*

However, even the largest terrestrial calderas are only a fraction of the diameter of many large lunar craters such as Bailly (311 kilometers), Clavius (231 kilometers), Copernicus (91 kilometers), and Tycho (87 kilometers). "Volcanists" argued that the lower lunar gravity made possible the formation of much larger volcanic collapse structures than could be formed on the earth, but the mechanical arguments in support of this idea were by no means convincing.

More importantly, the "impacters" could point to a steadily increasing amount of evidence for the impact of giant meteorites on the earth in the recent past. By 1950 scientists had identified about a dozen craters formed by such impacts within the last 100,000 years. The largest one, the famous Meteor Crater in Arizona, formed less than 50,000 years ago, is more than a kilometer in diameter. If the earth has been hit by about a dozen large meteorites in only the last 100,000 years, then in the 4.6 billion years since the earth formed, there was ample time for it to be struck by thousands of even larger objects. There are still many large asteroids that cross the earth's orbit today, and there might have been many more of them in the past.

There are far more impact craters visible on the moon, the "impacters" argued, because even the oldest craters

* There are also numerous volcanic eruptions on the earth that do not produce either volcanoes or calderas. These eruptions occur as floods of lava that pour out of large fissures in the earth and spread out for many kilometers around them. Such eruptions have built up flat plateaus of basalt layers in places like Iceland, India, and the northwest United States. The lavas that cover the lunar maria were apparently produced by this type of eruption. The mare lavas are not related to individual craters, and many large craters such as Copernicus and Tycho are much younger than the mare lavas. For this reason, the discovery that the lunar maria are covered by basalt lavas does not affect the debate over the origin of lunar craters.

could be preserved on the airless, waterless moon. By contrast, the ancient craters on earth would have been largely destroyed by erosion or buried under layers of younger sedimentary rocks.

Depending on where one looked, the earth thus provided support for either the impact or the volcanic view, and the telescopic observation of the moon itself did little to settle the controversy. The surface of the moon is a complex and diverse place, which offered evidence which fit either the impact *or* the volcanic theory.*

Many craters, like Copernicus (photo 2), show features that one would expect from large impacts. Such craters are circular, and they are surrounded by hummocky blankets of a material that was apparently blasted out of them by the impact of a large body. Many of these craters, especially Tycho and Copernicus, have a radial pattern of bright rays that have been interpreted as long streamers of material ejected from the crater and scattered for hundreds of kilometers across the moon.†

* A characteristic of lunar studies has been the ability of almost any theorist to explain a new observation in a way that supports his particular theory, so that a single result (e.g., the Ranger 7 pictures) can be invoked to support any number of contradictory theories. This phenomenon, which also exists in other fields of science, was remarked on by H. C. Urey early in the Apollo Program, and it has sometimes been given semi-formal status as "Urey's Law." See R. S. Lewis, *The Voyages of Apollo* (New York: Quadrangle, 1974), pp. 13, 57, 133.

† The formation of a large meteorite impact crater is far more complicated than the experiment conducted by Hooke (in which he dropped bullets into wet clay) would indicate. A body that strikes the earth or moon will be traveling at 10 to 20 kilometers per second. At such a high velocity, the body's kinetic energy of motion (equal to one-half the product of its mass times the square of its velocity, or $\frac{1}{2}mv^2$) is equal to several times the explosive energy of the same mass of TNT. As a result, the formation of a crater does not result primarily from the penetration of the projectile but from the sudden release of all the energy as shock waves that excavate the crater and shatter and melt the target rock. There are two important consequences. First, even small bodies make large craters. The projectile that formed Copernicus was probably about 2 kilometers in diameter

Other large craters, for instance Plato and Alphonsus, provide equally good indirect evidence for the volcanic theory. Plato (photo 3) is filled with the kind of dark material that we now know are volcanic lava flows. The crater Alphonsus, the landing target for Ranger 9, has a large central peak and smaller dark-haloed craters on its floor. Alphonsus strongly resembles many large calderas on the earth. The possible volcanic origin of Alphonsus was strengthened in 1958 when the Russian astronomer N. A. Kozyrev (1908–   ) detected what might be glowing volcanic gas coming from the central peak.

The elongate winding canyons on the moon, called "sinuous rilles" (photo 4), are another lunar feature that generated a variety of competing theories. One suggested that they were actual river beds, formed in the ancient past when the moon might briefly have had water. Another proposed that the rilles had been cut by gas-rich flows of volcanic ash produced by violent eruptions. A third theory, also volcanic, had it that the rilles were tunnels and channels through which rivers of molten lava had flowed out onto the maria from vents extending deep inside the moon.

Nor was the nature of the dark maria themselves unanimously agreed upon. They could not have been actual oceans; and many scientists believed that they were composed of flows of basalt lava. Other possible explanations were: that they were dark sediments which had been deposited in lunar seas that had briefly existed long ago; or that the maria were filled with layers of fine dust as much as several kilometers deep, into which, it was further argued, both astronauts and spacecraft would sink without a trace. A great many robot spacecraft were employed to finally

(a fairly common size for asteroids), but its impact released the energy of a million one-megaton hydrogen bombs and formed a crater over 90 kilometers across. Second, oblique impacts still make circular craters because the pattern of explosive energy release is not greatly affected by the angle at which the projectile strikes the surface.

disprove this last idea before men actually took off for the moon.

By 1957, when the Space Age was inaugurated with the launching of the Russian satellite Sputnik 1, theories about the moon had reached the limit imposed by earthbound observations. Any further advances in lunar study would require something new in the history of science: a massive, all-out effort to study the moon at close range, to put instruments on its surface, and to bring its rocks back to earth for analysis.

Support for space exploration and the lunar landings came from many fields of science, each interested in different basic questions. Geologists, unable to discover the early history of our own planet, looked to the moon as a place where the primordial record might have been preserved. Astronomers interested in the origin of the solar system also wanted more information from the moon. For them, the moon was a possible site for an observation platform above the earth's murky atmosphere—even a small telescope on the moon could see as much as the 200-inch (508-centimeter) Mount Palomar telescope on the earth. In their turn, physicists wanted to use the moon as a base for direct observation of the matter and energy that pass through space, since they had heretofore been able to observe only the by-products of reactions with our protecting atmosphere. And biologists, who are always interested in the origin of life on earth and the possibility of life elsewhere, wondered if the moon supported life now, whether it might ever have supported life in the past, or whether its rocks contained substances that, given time and a more propitious environment, might have developed into life.

Scientists could speak easily of the new knowledge to be gained by going to the moon, but no one could visualize what the decision to go to the moon would involve. Sending men to the moon would be difficult, complicated, probably dangerous, and certainly expensive. There would be no miraculous propulsion systems, no brilliant but poverty-stricken scientists building spaceships in basements and back yards. The Apollo Program would require the combined resources of government, science, and technology, mobilized on a scale that would dwarf the building of the pyramids and the great cathedrals of the Middle Ages. And the American public and their elected representatives would have to become enthusiastic enough about it to pay the bill.

In the course of the next decade, the Apollo Program developed into one of the most impressive human efforts that man has ever produced. It was responsible for massive buildings and the development of great cities in Texas swamplands and along Florida beaches. It drew on the scientific resources of 120 universities and the technical abilities of 20,000 American firms. At its peak it directly employed half a million people and indirectly provided work for several million more. Its chief product, the Apollo/Saturn spacecraft, was over 100 meters tall, weighed over 2½ million kilograms at launch, and contained 15 million separate parts.

It was hard to be neutral about going to the moon. Both the philosophical implications and the economic resources involved were too great to be ignored. People were either very much for the Apollo Program or very much against it. It would be hailed as the greatest event since man's ancestors crawled out of the ocean onto dry land, but at the same time it would be denounced as a useless technological stunt and derided because there was nothing either philosophical or poetic about it.

The program would advance American technical know-how in a way never before achieved without the pressure of a major war. It would produce improved computers, new medical instruments, refined techniques for management and production, yet it would also be denounced because there were no "practical results."

Although hundreds of thousands—perhaps millions—of people were employed by the Apollo Program, it would be criticized because the extensive budget funding it received might have been used for the delivery of social services.

The Apollo Program cost the American people less than one-fifth of what they spent on cigarettes and liquor alone during the same period, but many of its critics would say that we couldn't afford it.*

Money would not be the only cost however. Three astronauts would die in a spacecraft fire on the ground, and three more would be killed in plane crashes while training. Hundreds of anonymous young men and women would spend their most productive years in frustration, mapping landing sites that were never visited, building equipment that was never used, and training for missions that never flew. Thousands more would work in isolation and near-frenzy, perhaps never even seeing the whole sweep of the program, doing the millions of small tasks that had to be done on time and had to be done right, yet continually aware of how little margin the universe allows for ignorance or error.

* The cost of the Apollo Program has been estimated at between 25 and 40 billion dollars for the decade 1960–70. For an American population of about 200 million people, the annual cost of the Apollo Program was less than $20 per person. During this same decade, the average American also spent each year over $80 on cigarettes (about 4,000), over $50 on beer (about 95 liters), and over $50 on liquor (about 8 liters). (Data from *Statistical Abstract of the United States* [U.S. Bureau of the Census (Washington, D.C., 1971)] pp. 374, 701–2.)

CHAPTER

THE LOOK BEFORE THE LEAP

"I believe that this nation should commit itself to achieving the goal, before this decade is out, of landing a man on the moon and returning him safely to earth. No single space project in this period will be more impressive to mankind or more important for the long-range exploration of space. And none will be so difficult or expensive to accomplish."

The speaker was President John F. Kennedy; the occasion was the announcement to Congress on May 25, 1961, of his decision to implement the Apollo Program. On that day, man's long curiosity about the moon became part of a major national effort.

There were many reasons for undertaking the Apollo Program in 1961. It would restore American prestige, which had been diminished by Russian space successes and by the recent and disastrous failure of American-supported Cuban exiles at the Bay of Pigs. It would represent a new boost for American technology; and it would provide jobs in a troubled economy. Because it had never been done, the idea of going to the moon was a challenge to a nation that took pride in its ability to do the impossible. And it would provide new information about the universe that could never be obtained without leaving the earth. But although scientific goals were only one of many forces that brought the Apollo Program into being, scientific information was one of its most important results.

The long-debated questions about the nature, origin, and environment of the moon suddenly acquired very real and very practical importance. Engineers faced with the task of building real spaceships to carry men safely to the moon, land them, and bring them safely back, needed definite answers to extremely difficult questions. Was the lunar surface covered with a fine dust hundreds of meters thick that would swallow spacecraft and astronauts together? Could a man survive on the moon's surface? Could he move around and work? What were the chances of his being hit by a meteorite? How strong should his spacesuit be to protect him while still allowing him to move around? Were lunar rocks made up of strange materials that might burst into flame when brought into the oxygen atmosphere of the spacecraft?

Scientists had their own questions to answer too: Where should the astronauts land to obtain the most information about the moon? What kind of observations should they make? What kinds of samples should they collect? What instruments should be set up on the moon to transmit back data after the astronauts had left?

At the start of the Apollo Program, there were a multitude of these unanswered questions. Despite centuries of study, we still did not know very much about the moon's surface. The best telescopes on earth could only resolve objects at least a kilometer wide on the lunar surface. Throughout the 1960s scientists were engaged in a major effort to obtain as much information as possible about the moon before trying to send men there. One activity involved the intensive study of the moon through earth-based telescopes. Another series of programs sent robot spacecraft off to the moon—first Rangers and then Orbiters to photograph its surface, along with Surveyors to land gently and test the footing for the men who would follow.

The results of this widespread collective effort would determine whether a manned mission was possible or whether the idea would have to be abandoned as too dangerous. These projects were also intended to determine what equipment would be needed, what kind of training the astronauts would require, and where they should land on the moon.

While these studies were being carried out, scientists were also studying the earth's volcanoes and meteorite impact craters in great detail in order to compare them with similar features on the moon. As this work went on, astronauts worked side by side with geologists, learning to observe, describe, and collect rock samples, practicing on the earth the work that they might do one day on the moon.

TELESCOPIC MAPPING

The moon and the planets had been largely ignored by astronomers during the first half of the twentieth century. There were several reasons. For one, the moon had been as well mapped as was possible with existing telescopes, and the planets were even more difficult to observe. In addition, the revolutionary discoveries in spectroscopy and atomic physics had opened up exciting possibilities for studying the composition and motions of distant stars and galaxies, so that when large new telescopes were built, they were aimed at the distant universe instead of at the more familiar moon. And finally, the image of planetary astronomy had just been tarnished by the widely publicized controversy about life on Mars.

This argument began innocently enough in 1877 when the Italian astronomer Giovanni Schiaparelli (1835–1910) reported the existence of a network of lines on the surface of Mars. Schiaparelli did not suggest that the features, which he called *canali*, were anything but natural, but the word *canali* (meaning "channels") was somewhat freely translated by others as "canals," with all the implications of artificial construction. Several astronomers, most notably the American Percival Lowell (1855–1916), viewed the canals as a great irrigation system devised by intelligent Martians to carry water from the polar caps toward the equator. Other astronomers argued heatedly that the channels were either natural, or were an optical illusion, or just weren't there at all. Because observing Mars pushed existing telescopes to their limits, the results tended to be subjective—either strongly influenced by small changes in the earth's atmosphere or by individual quirks in the eye and mind of the viewer. No two observers saw the same thing, and different estimates of the number of "canals" ranged from zero to upwards of several hundred.

The debate eventually subsided, but was not definitely settled until Mariner spacecraft observed Mars from close range. Close observations of Mars by unmanned spacecraft have revealed no "canals" whatsoever. Photographs taken during flybys of Mars in 1965 (Mariner 4) and 1969 (Mariners 6 and 7), and extensive photography by the Mars-orbiting Mariner 9 (1971) show craters, volcanoes, canyons, and wind-produced surface streaks, but it now seems clear that the intricate network of lines described by Schiaparelli and Lowell never existed. Most likely they were an optical illusion produced by the eye's tendency to perceive connecting lines between more or less random spots and blotches—such as the ones that actually mark the surface of Mars. In other words (to paraphrase one astronomer) the canals *were* produced by intelligent beings, but the intelligence was on the other end of the telescope.

Unfortunately for us, in the early part of the century when the dispute over the canals was at its height, its notoriety caused many young astronomers to avoid the apparently controversial and undignified area of planetary astronomy and move on to the more exciting action available in the stars and galaxies. For nearly half a century, astronomy bypassed our own solar system in its outward rush to the far corners of the universe.

Nevertheless a few astronomers continued to map, study, and speculate about the moon in the pre-Apollo years. One was astrophysicist Ralph B. Baldwin (1912–), who presented in two books, *The Face of the Moon* (1948) and *The Measure of the Moon* (1963), a synthesis of lunar observations, theoretical studies, and geological information that represented a major contribution to our understanding of the origin and evolution of the moon.

Baldwin presented evidence for the impact origin of lunar craters, drawing close comparisons between them and the "fossil meteorite craters" just then being recognized on the earth. He argued that the moon was an ancient world whose heavily cratered highlands preserved the record of an intense primordial bombardment that might have been the last event in the formation of the moon itself. He suggested that the moon was an evolved planet that had separated chemically to produce different kinds of rocks. And he proposed that the dark maria were composed of layers of basalt which had poured out into basins formed earlier by the catastrophic impacts of huge asteroids tens of kilometers in diameter.

Baldwin's lunar work was combined with a full-time career as vice-president of the Oliver Machinery Company of Grand Rapids, Michigan. Although he never became prominent in the Apollo Program, he saw his ideas "discovered" again and largely confirmed as each Apollo mission brought back new information from the moon.

Another pre-Apollo landmark was the publication of *The*

Planets, written in 1952 by H. C. Urey, who won the Nobel Prize in 1934 for his discovery of "heavy" hydrogen. Urey examined the solar system from a theoretical viewpoint, arguing that the laws of physics and chemistry provided a framework for understanding how the planets had formed and evolved. Suddenly, now that they were objects whose history might be accurately deciphered, the planets became interesting again. The moon became especially important, for Urey suggested that it was a primordial object, cold and unchanged since the solar system had formed, a Rosetta Stone for understanding the rest of the solar system.

With the birth of the Apollo Program, telescopes were again turned toward the moon for advanced geologic surface mapping. Much of this work was carried out by geologists from the U.S. Geological Survey's Branch of Astrogeology in Flagstaff, Arizona. These geologists began to observe the moon through Lowell's old 24-inch (38-centimeter) refracting telescope—the same one that had figured so prominently in the Mars controversy over half a century earlier. After a modern 30-inch (76-centimeter) reflector was added in 1964, the mapping program speeded up.

The foundation for this work had been established by E. M. Shoemaker and his colleagues of the U.S. Geological Survey. In a brief article published in 1962, formidably entitled "Stratigraphic Basis for a Lunar Time Scale," Shoemaker argued that the principles already used for terrestrial geology could be applied, with some modifications, to the moon.*

Those principles were simple enough, and geologists proceeded to apply them. It had been obvious from the first that the moon was divided into two main regions: the lighter, rugged, and heavily cratered highlands, and the darker, flatter, and less cratered maria. But with more care-

* E. M. Shoemaker and R. J. Hackman, "Stratigraphic Basis for a Lunar Time Scale," in *The Moon* (New York: Academic Press, 1962), pp. 289–300.

ful use of the telescope it became apparent that both regions contain many different rock units, each of which is distinctly different in appearance, in color, in the amount of light it reflects, and in the number and size of craters formed on it. A geological map of the visible side of the moon, made from these observations, divided the moon's surface into about 20 different rock units, even though the chemical composition and origin of these rocks remained uncertain.

Once these different rock formations were identified, the geologists tried to distinguish the older from the younger ones, so that they could assemble the individual units to form a history of the moon. Although telescopic observations alone cannot determine the precise ages of lunar rocks they can be used to estimate *relative* ages of different rock units. A basic rule of both terrestrial and lunar geology is that younger layers of rock are deposited on top of older ones. For example, the bright rays of the crater Copernicus (see photo 2) have been deposited over and around the more subdued crater Eratosthenes. The relation shows that Copernicus is the younger of the two craters. Another simple application of this geological principle on the moon is illustrated by the many craters around the maria which have been flooded or partly buried by the dark mare material. In these cases, the craters must be older than the dark material which fills them.

Some geologists believe that nearly all the lunar craters were formed by meteorite impact. In their view, the moon is being continually bombarded by many particles ranging from tiny bodies smaller than a pinhead to asteroids several kilometers in diameter. The smaller bodies are more numerous and strike the moon almost continuously; the larger ones are much rarer and hit the moon only once every few million years.

One effect of this continuous bombardment by small particles is to erode the sharp outlines of freshly formed larger craters and gradually to destroy the bright rays that originally surrounded them. As this process goes on, older craters become more blurred, indistinct, and subdued. Craters such as Eratosthenes, which are relatively subdued and lack a visible ray pattern, are therefore thought to be older than the sharp, fresh, rayed craters like Copernicus and Tycho.

This meteorite bombardment also provides a unique way to measure the relative ages of rock units and craters on the moon. The longer a rock unit is exposed to continuing impacts at the lunar surface, the more particles will hit it and the more craters will be formed on it. Thus, older formations on the moon will have more craters than younger ones. Therefore the heavily cratered highlands are interpreted as being significantly older than the less cratered maria. More detailed subdivisions of the moon could be made with crater counts, and formations with different apparent ages could be distinguished even within the highlands and maria themselves.

On the other side of the controversy, scientists who favor a volcanic origin for the lunar craters argue that time scales cannot accurately be based on crater counts because the craters might have formed in isolated episodes of volcanic action rather than by a continuous process operating over the whole lunar surface. To counter this argument, the "impacters" point out that the general consistency of their crater-count ages with other geological observations is supporting evidence for the impact theory. In any case, crater counts are objective data which can be used to test both the impact and volcanic theories as soon as the ages of returned lunar rocks are measured. If the ages calculated by the "impact" theorists should turn out to be seriously wrong or inconsistent, then the impact theory itself would be in trouble.

The large lunar craters, however, provide a way of timekeeping that most geologists, both "impacters" and "volcanists," could accept. It is clear that however the large craters and even larger mare basins* originated, their formation was a sudden and catastrophic event during which a great deal of material was blasted out of the crater and scattered across the surface of the moon all at once. The excavated material (or *ejecta*) from large craters may be deposited as a separate recognizable layer for hundreds of kilometers around the crater; such a layer can serve as a unique record of a single episode over wide areas of the moon.

Using these principles, geologists produced surface maps and suggested possible landing sites for the projected manned space missions. For example: the map for the Hadley Rille region, published by the U.S. Geological Survey in 1966, described the geology that would be sampled by the Apollo 15 astronauts five years later.†

Hadley Rille and the nearby Apennine Mountains are located on the rim of Mare Imbrium, which is the dark region 1,600 kilometers in diameter that constitutes the right eye of The Man in the Moon. Many lunar scientists, among them Baldwin, Urey, and Shoemaker, have proposed that the basin of Mare Imbrium was formed when a huge asteroid about 100 kilometers across struck the moon. This catastrophe formed a huge crater several hundred kilometers in diameter (the Imbrium Basin) and scattered a thick blanket of material, now called the Fra Mauro Forma-

* An important distinction is made between the lunar *maria* and the *mare basins* which contain them. The term *maria* is used to designate areas of the moon covered by a dark "mare material" which has now been identified as flows of basalt lava. A *mare basin* is the depression, perhaps 10–20 kilometers deep and hundreds of kilometers across, that contains the dark mare material. The formation of the mare basin occurred first, and the basin was filled later with dark lava flows to form the mare itself. Thus, *Mare Imbrium* designates the dark circular region on the moon, whereas the *Imbrium Basin* refers to the great circular depression that contains it.

† R. J. Hackman, *Geologic Map of the Montes Apenninus Region of the Moon* (U.S. Geological Survey Map I–463 (LAC–41), 1966).

tion, over most of the earth-facing side of the moon. Because the effects of this so-called Imbrium Event can be recognized over so much of the moon, the Imbrium Event is used as the "Year Zero" in lunar history; all other lunar rocks are dated as pre-Imbrium or post-Imbrium.

In the geologic map of the Apollo 15 landing site (photo 5) the edge of Mare Imbrium runs from the upper right to lower left, and the surface of the mare stretches away to the upper left. With a telescope, the geologists distinguished four different formations and surmised their relative ages. The oldest rocks, designated as pre-Imbrium, form high, rugged mountains (the Apennines) which rise just outside Mare Imbrium (in the lower right of the map). The mountains are believed to have been shoved up very suddenly as the force of the same impact formed the Imbrium Basin.

After the catastrophic Imbrium Event, a number of other fairly large craters were formed in this region. One of them, Archimedes, is the large crater at the upper left of the map. Around Archimedes is a halo of material (colored dark gray on the map) that was ejected when the crater formed. This layer, or "ejecta blanket," must have originally extended all around Archimedes, but part of it is now covered by the rocks (shown in lighter colors on the map) that have since filled the Imbrium Basin.

We know that Archimedes must be *younger* than the Imbrium Event itself because otherwise it would have been destroyed in the catastrophe. Similarly, Archimedes must be *older* than the rocks that cover the ejecta blanket around the crater and flood the crater itself.

These observations tell us something very important about the history of the moon: The formation of the huge Imbrium Basin and the flooding of the basin by the dark mare material could not have taken place all at once. Enough time to form large craters like Archimedes must have elapsed after the Imbrium Event before both the Imbrium Basin and these younger craters were covered with the dark rocks that now make up Mare Imbrium.

Telescopic observations alone could not ascertain when the various features formed or how much time passed between the Imbrium Event and later developments. But the geological history outlined for the Hadley Rille region from earth-based observations was strikingly confirmed when lunar samples were returned from Apollo 14 and 15. In 1971, the Apollo 14 mission sampled the Fra Mauro Formation, located in another part of the moon. From these samples geologists determined that the Imbrium Event occurred about 3.9–4.0 billion years ago. Apollo 15 samples from Hadley Rille included the dark rocks that cover Mare Imbrium and the Archimedes crater. These rocks were identified as basalt lava flows and dated at 3.2–3.3 billion years old. These data indicate that the hiatus between the formation of the Imbrium Basin and the later lava flows into Mare Imbrium was about 700 million years—nearly ten times the span of time between the dinosaurs and man.

The map of the Apollo 15 landing site also shows numerous elongate, winding canyons ("sinuous rilles") that cut across the surface of Mare Imbrium. The basic rules of geology indicate that the canyons must have formed later than the solid rocks they have been carved into, but the exact age of the canyons themselves is not known. The results from the Apollo 15 mission suggest that these rilles are feeder channels through which lava flowed on its way onto the surface of Mare Imbrium. If this interpretation is correct, then the rilles must have formed during the filling of Mare Imbrium, and both the rilles and the lava flows have practically the same age, i.e., 3.2–3.3 billion years.

The youngest features on the map are scattered craters that have formed on both the older pre-Imbrium rocks and the younger lava flows in Mare Imbrium. These craters are identified as the most recently formed because each is surrounded by a halo of ejected material, and neither the craters nor their halos have been covered up by any younger rocks. The craters probably came into being at various times during the last three billion years. Their exact ages remain unknown.

ROBOT SPACECRAFT

While the telescope was being used to map the geology of the moon and piece together its history, three different kinds of unmanned spacecraft were launched to also give us a closer look at the moon and to check for hazards at the possible landing sites.

Ranger

In the late 1950s, the Russians were clearly ahead of the United States in sending unmanned spacecraft to the moon. In 1959, they successfully impacted two spacecraft, Luna 1 and Luna 2, onto the moon. Shortly afterwards, in October, 1959, a photographic spacecraft, Luna 3, circled the moon and radioed back the first views of its far side ever seen by man.

The Luna 3 pictures were a major step forward in the examination of the moon, and they provided an important impetus for further exploration. Although their quality was poor compared with the pictures we obtained later, they clearly showed the surface of almost all of the far side of the moon to be a bright, heavily cratered terrain similar to the highlands on the earth-facing side. There were many large craters, but only a few patches of dark mare material. Some maria and craters on the far side of the moon were seen clearly enough to be given names, particularly Mare Moscoviense (the Sea of Moscow) and the crater Tsiolkovsky.

Against these Russian successes, the American program to

explore the moon was going badly, with one frustrating failure after another. Three attempts to launch spacecraft into orbit around the moon failed completely in 1959 and 1960 before the Ranger Program was begun in 1961.*

The Ranger spacecraft were built to provide better close-up views of the lunar surface than could be obtained from earth-based telescopes. The spacecraft were intended to fly directly to the moon and to transmit back close-up TV pictures of its surface before they crashed into it at full speed. At first, the program suffered a series of discouraging failures. After Rangers 1 and 2 were tested in earth orbit in 1961, Ranger 3 was launched toward the moon in January, 1962. Unexpectedly, its third-stage rocket failed, causing it to miss the moon. Three months later, Ranger 4 was launched, but radio control was lost and it crashed into the far side of the moon, thus becoming the first American object to hit the lunar surface. In October, 1962, Ranger 5 suffered a power loss while en route and shot past the moon into orbit around the sun. In January, 1964, Ranger 6 made a perfect crash landing on the moon only 32 kilometers from the intended target, but the TV camera had stopped transmitting even before the spacecraft reached the moon, so no pictures were returned.

Although the Ranger Program began with failure, it closed with three astonishing successes. In July, 1964, Ranger 7 crashed only 16 kilometers away from the target point in *Mare Nubium* (the Sea of Clouds). The TV cameras worked perfectly; 4,316 close-up photographs of the moon were returned to earth, and ecstatic scientists promptly named the "landing" site *Mare Cognitum* (the Known Sea). In February, 1965, Ranger 8 returned 7,137 pictures before crashing into the surface of Mare Tranquillitatis only 23 kilometers away from its target point (and 68

* These early attempts, made with small Pioneer spacecraft, should not be confused with the highly successful Lunar Orbiter program carried out later in 1966–68.

kilometers from what, with the landing of Apollo 11, would become Tranquillity Base). Only one month later, Ranger 9 scored a perfect bull's-eye in the crater Alphonsus, radioing back 5,814 pictures before it crashed less than 5 kilometers from its intended site near the crater's central peak.

The Ranger photographs increased our knowledge of the surface of the moon by a factor of 1,000. Now boulders and small craters less than a meter wide were visible to us. On this small scale, the surface of the moon appeared smooth and gently rolling; there were no cliffs, ledges, or large blocks of bare rock. The surface appeared to be a layer of rubble that covered whatever lunar bedrock lay beneath it.

What surprised scientists most about the Ranger pictures was that they revealed a moon that was much more heavily cratered than had appeared from earth, with thousands of craters ranging in size from a few hundred meters across to barely visible pits less than a meter wide. The smaller craters seemed to many scientists to provide supporting evidence for continuous bombardment of the moon, because they were just the right size to have been formed by the walnut- to grapefruit-size meteorites which produce "shooting stars" when they strike the earth's atmosphere and burn up.

Ranger 9, the last of the Ranger missions, captured some spectacular scenic views when it plunged into the crater Alphonsus. The floor of Alphonsus was pockmarked with thousands of small craters, indicating that bombardment by small particles had taken place in this area of the moon as well. Furthermore, Ranger 9 also took some intriguing close-ups of possible volcanic features within the crater, including a number of small, dark-haloed craters from which volcanic gases might have come.

Before the first Ranger was launched, scientists had hoped that the first close-up pictures of the moon would settle all the major arguments about it. It didn't work out that way. "Urey's Law" continued to operate, and each scientist found in the new images some features that supported his own particular view of what the moon was like. But Ranger did provide a great deal of concrete and important information. It showed that the moon was heavily cratered and that much of the lunar surface was a smooth, rolling, pockmarked plain. And most important, it showed that if the surface were strong enough to support it, a spacecraft *could* land on the moon.

Surveyor

The next step, then, was to build a spacecraft that could be landed intact to take photographs and determine the physical properties of the surface layer. A "soft-lander" was a far more complicated spacecraft than the Rangers. It carried a powerful rocket motor that was fired as the spacecraft approached the moon, slowing the spacecraft so that it could land gently on the lunar surface.

At this stage of lunar exploration, the United States still lagged behind the Soviet Union. Russia's Luna 9 landed safely on the moon on February 3, 1966, carrying an instrument package shaped like a metal ball which unfolded like the petals of a flower, propped up a camera, and aimed it at the surrounding moonscape. The pictures it sent back showed a view of what was apparently a field of rocky rubble. This was important because Luna 9 sat firmly on the surface and showed no tendency to sink and vanish from sight—proof that in at least one area of the moon, the surface layer was strong enough to support a machine.

The United States space program caught up quickly. On June 2, 1966, the first spacecraft in the new Surveyor Program dropped gently to the surface of Oceanus Procellarum (the Ocean of Storms), swung its TV camera into action, and began to transmit back to earth pictures of the lunar surface on which it rested solidly. In those first few minutes,

Surveyor 1 fulfilled the major goals of its mission. It established that the lunar surface would support a heavy machine—or a man. The Surveyor 1 landing indicated that the lunar surface would support loads of about 350 grams per square centimeter; the surface layer was about as strong as wet beach sand or ploughed farm soil, and men and machines could travel safely across it. The plans for landing

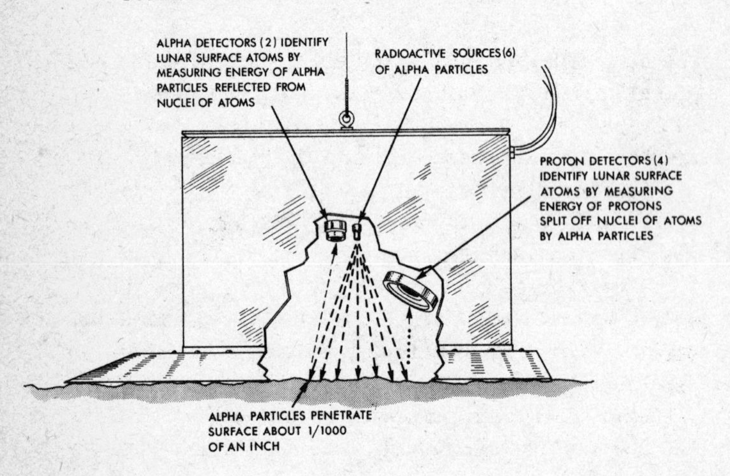

Figure B / The Analysis of the Moon. The equipment used by the Surveyor spacecraft to obtain chemical analyses of the lunar surface is shown in this cutaway drawing. Inside a gold-plated box, which is lowered to the lunar surface on a cable, are six radioactive sources that spray the lunar surface with alpha-particles (helium atoms). The alpha-particles strike atoms in the lunar soil and bounce back to detectors in the box. Other detectors record protons (hydrogen atoms) that are knocked out of the lunar soil atoms by the alpha-particles. Data from the two sets of detectors make it possible to calculate the chemical composition of the lunar soil beneath the instrument. (Based on NASA photograph 67–H–1184)

men on the moon went ahead with a new burst of confidence. Surveyor 1 pictures also showed an interesting site with rocks and boulders lying around waiting to be collected. By the time the camera power was exhausted, Surveyor 1 had transmitted 11,237 pictures, in which appeared boulders up to a meter in length and pebbles as small as a few millimeters across.

The Surveyor Program was far more successful than the Ranger Program had been. Seven spacecraft were launched, and five landed safely on the moon. Together they returned nearly 100,000 close-up photos of the lunar surface. The Surveyor pictures provided another 1,000-fold increase in resolution over the Ranger pictures; they clearly showed small craters, pebbles, and clods of lunar soil as small as a millimeter in size. Some of the Surveyors dug trenches and prodded the lunar soil with movable scoops, confirming the strength of the ground and exposing to the TV cameras the nature of the first few centimeters of soil beneath the surface.

Surveyor 5 landed safely on the surface of Mare Tranquillitatis on September 11, 1967. Shortly after landing, radio commands from earth caused the spacecraft to lower a small gold-plated box to the ground to begin the first chemical analysis of the surface of another planet. Inside the box was a tiny amount of an artificial radioactive element that emitted small alpha-particles.* As the tiny alpha-particles shot out and struck the much larger atoms in the lunar soil beneath the box, they were scattered in all directions, just as a stream of water splashes when it hits a pile of loose rocks. Some of the alpha-particles were scattered back into the box where a detector recorded them (*Figure B*). The heavier

* An *alpha-particle*, composed of two protons and two neutrons, is a nucleus of a helium atom. Alpha-particles are common products of the decay of natural and artificial radioactive elements such as thorium, uranium, and plutonium. The alpha-particles for the Surveyor 5 experiment were obtained from an artificially produced radioactive element, curium-242.

the atoms in the lunar soil, the more alpha-particles were scattered backward. From the number of alpha-particles scattered back into the box, scientists calculated the weights of the atoms that made up the lunar soil, and from their weights, the elements themselves could be identified.

After 900 hours of this experiment, enough data had been gathered to come to some conclusion about the composition of Mare Tranquillitatis. The soil from this area appeared to be more like terrestrial basalt than any other type of rock or meteorite. The chemical composition of the soil broke down as follows: oxygen, 58 percent; silicon, 19 percent; aluminum, 7 percent; magnesium, 3 percent; plus 13 percent of heavier elements which this experiment could not resolve (chiefly iron and titanium, as it turned out). These first chemical analyses of the moon revealed that its rocks were very much like the earth's. The results of the Surveyor 5 experiment (later confirmed by laboratory analysis of the Apollo 11 samples collected two years later, in 1969) dealt a fatal blow to the idea that the moon was an unchanged primitive object, with a totally different composition from the earth.

Chemical analyses made by Surveyor 6 in another mare region, Sinus Medii, also indicated a basalt-like composition. Surveyor 7, whose destination was the lunar highlands, touched down near the crater Tycho; the soil which it analyzed was part of the blanket of material ejected from the crater. In this highland material, scientists found more aluminum and less heavy metals like iron and titanium than had been found in the maria.

The geological findings of the Surveyor program were landmarks. The Surveyor spacecraft provided the first direct evidence of a chemical kinship between the moon and the earth. By confirming that the maria soils were similar to terrestrial basalts, the Surveyor experiments proved that the moon was not related to the primitive meteorites; they strengthened the opposing view that it was an evolved planet with a history of chemical change. The program also showed that the maria and highlands were chemically different, implying that the moon was not a uniform planet. The engineering results of Surveyor were equally important. The safe landings, the pictures, and the trenches dug into the lunar soil dispelled doubts that the lunar surface might not support a manned spacecraft.

Orbiter

At the same time that the Surveyor spacecraft were prodding and analyzing the lunar soil, other spacecraft were orbiting the moon in order to carry out a photographic survey of its surface. The goal of the Lunar Orbiter Program was to provide high-resolution photographs of several sites in the lunar maria where the first manned landing might be made. These sharp, high-resolution photographs were detailed enough to show boulders and craters only a few meters across; they were scanned by both scientists and engineers for hazards and for important geological features.

Although the Russians had continued photographing the moon since the voyage of Luna 3 in 1959 with other spacecraft, this early work was superseded by the flood of high-quality, high-resolution pictures transmitted back by the Orbiters. The program was one of the most successful in the history of space exploration; five spacecraft were launched, and all five worked perfectly. The program started with the launch of Orbiter 1 on August 10, 1966, and ended on January 31, 1968, when Orbiter 5, its mission completed, was crashed into the moon to remove it from the path of the Apollo missions to come. The results from the Orbiter Program in those 18 months far exceeded all the planners' hopes and expectations.

The original aim of the project, the close-up survey of eight possible landing sites, was completed during the first three Orbiter missions. This efficiency allowed Orbiters 4 and 5 to carry out lunar mapping and other scientific observations of the rest of the moon.

More than 95 percent of the surface of the moon was photographed by Orbiter spacecraft. The thousands of pictures they returned could fill a huge room; they provided enough information to keep scientists busy for many years.

Some of the Orbiter pictures were spectacular enough to change the way people thought about the moon. From a vantage point a few hundred kilometers above the moon, the Orbiter cameras showed wide expanses of the lunar surface that merged in the distance with the curve of the lunar horizon. It was hard to look at those pictures and still think of the moon as a distant and incomprehensible object in the sky. Many of these Orbiter panoramas have become classic portraiture of the Space Age. One photograph captures the wide sweep of Oceanus Procellarum with its flat, level plains and the irregular "wrinkle ridges." Another presents a low-angle distant view of the crater Copernicus with the curious keyhole-shaped crater Fauth in the foreground. And in what was called "the picture of the century," Orbiter 2 looks at the crater Copernicus from just above its rim (photo 6), revealing the central peak with its terraced walls twice as high as the Grand Canyon.

When the Orbiters swung repeatedly around the far side of the moon, they sent back sharp, clear images of things that had never been seen before. Mare Orientale (the Eastern Sea), barely visible from the earth, straddles the boundary between the near and far sides of the moon. In Orbiter photographs, with their new viewpoint, Mare Orientale appears as a huge, multi-ringed bull's-eye about 900 kilometers in diameter. Geological studies indicate that this structure is the youngest of the large mare basins, even younger than Mare Imbrium, whose formation is the chief marker in the lunar time scale. Curiously, although the Orientale structure is almost as large as the Imbrium Basin, it contains practically no dark lava except for a few patchy areas near the center. Why Mare Imbrium and the other mare basins were flooded with lava, and Mare Orientale was not, is a question that is still unanswered.

Examining the Orbiter photographs, scientists discovered that there are large circular basins like Imbrium and Orientale on both sides of the moon. In fact, about as many large basins were discovered on the far side of the moon as on the near, or earth-facing, side. However, the far-side basins contain almost none of the dark lava that flooded the basins on the near side. This dark mare material is confined almost entirely to the near side of the moon. The far side of the moon, revealed in detail by the Orbiters, consists of heavily cratered highlands with only a few scattered areas of dark material. One dark patch in the crater Tsiolkovsky (photo 7) is about 250 kilometers across. Tsiolkovsky may be an example of Mare Imbrium in miniature—a large circular crater formed by a meteorite impact and later filled with lava flows.

The Orbiter pictures also helped fuel the controversy about meteorite impact and volcanism on the moon. Many of the newly photographed craters certainly were produced by impact, but other pictures show features that look volcanic. The Marius Hills near the large crater Marius (photo 8) are seen as a group of low, rounded, volcanic domes with possible lava flows—very unlike anything formed by meteorite impact. The visual evidence for relatively recent volcanism in the Marius Hills is so impressive that the area was considered as a landing site for Apollo 17.

As often happens in science, one of the major results of the Orbiter Program was unintentional and unexpected. The speed and location of all five Orbiters as they circled the moon were measured precisely and continuously by scientists on earth, using radio tracking equipment. Careful analysis of the data indicated—quite surprisingly—that the spacecraft did not move uniformly around the moon. Their average speed was several thousand kilometers an hour, but sometimes they speeded up unexpectedly, and at other times they slowed down. These changes in speed were slight (only a few meters per second, or about 0.1 percent of the orbital velocity) but they were large enough to arouse curiosity.

Eventually, a pattern was detected in the variations. Each time an Orbiter approached one of the dark circular maria on the near side of the moon, its speed increased; as it passed over the mare, it slowed down. Scientists concluded that there was some extra mass under the maria that made the moon's gravitational field a little stronger, thus producing variations in velocity of the spacecraft. This extra mass under the maria pulled the Orbiter toward the maria as it approached, then held it back a little as it passed over.

The unforeseen discovery of these concentrations of extra mass (the term was abbreviated to *mascons*) provoked a great deal of discussion. Eventually twelve such mascons were found on the near side of the moon. (The existence of mascons on the far side could not be firmly established because the Orbiter spacecraft could not be tracked when they were behind the moon.) Each mascon was associated with a dark circular mare such as Mare Imbrium, Mare Serenitatis, or Mare Crisium; no mascon was identified with the broad, irregular expanse of dark lava flows that comprises Oceanus Procellarum.

The existence of mascons indicated that the rocks under the maria must be denser and heavier than the rocks around them. Scientists were surprised to find excess mass under the maria, which were the lowest and flattest places on the moon, and several theories were offered to explain the phenomenon. One explanation was that the extra mass came from the remnants of giant asteroids that had crashed into the moon to form the mare basins. Other scientists pointed out that the extra mass could have resulted from a layer of basalt about 5 to 10 kilometers thick in the maria, but only if the basalt was about 10 percent denser than the rocks which formed the surrounding lunar highlands. This second theory would later be strongly supported by the nature and properties of the Apollo samples themselves.

Regardless of the exact nature of the mascons, their very existence indicated something important about the interior of the moon. Everyone agreed that the mascons were near-surface features, and that they were as old as the maria. This meant that the lunar interior was strong enough to have supported the extra mass for a long time, probably for a few billion years. If the moon's interior were weak and plastic, the excess mass would have sunk slowly into the moon and would not have been detectable by spacecraft passing over it.

To many scientists, the continuing existence of mascons implied a strong and rigid lunar interior. Because rocks and other solid materials lose their strength as they are heated, a rigid lunar interior meant a relatively cold lunar interior as well. The temperatures inside the moon must now be no more than a few hundred degrees Centigrade—far below the melting point of the lunar rocks. The Lunar Orbiters, which had been intended to map only the surface of the moon, had unexpectedly provided some important information about the moon's interior as well.

Ten years of planning and effort came to their climax within a few months in 1968 and 1969. Apollo 8, launched on December 21, 1968, carried Frank Borman, Jim Lovell, and William Anders to the moon and into orbit around it.*

* Apollo 1, 2, and 3 were unmanned missions in 1966 that tested separate components of the complete Apollo spacecraft system. Apollo 4, 5, and 6, in 1967 and early 1968, were more complicated unmanned tests. Apollo 7, launched on October 11, 1968, was the first manned Apollo mission, carrying three astronauts and testing the spacecraft system in orbit around the earth. Apollo 8 was the first Apollo mission actually launched toward the moon.

They became the first human beings to enter the gravity field of another planet and the first to see the earth rise above the surface of another world. Their trip provided a memorable Christmas Eve, as TV pictures of the moon's surface were relayed to earth while the astronauts read from the Book of Genesis to their listeners a quarter of a million miles away.

Three months later, in March, 1969, James McDivitt, David Scott, and Russell Schweickart, in Apollo 9, tested the complete Apollo system in orbit around the earth. In May, 1969, Apollo 10 carried Eugene Cernan, John Young, and Thomas Stafford to the moon for a "dress rehearsal" that involved every part of the lunar mission except the actual landing. Cernan and Stafford flew their Lunar Module closer to the moon than any men had ever been, flying as near to the lunar surface as airplanes fly above the earth. The success of Apollo 10 dispelled the last uncertainties about a lunar landing. Even before the astronauts returned to earth, Apollo 11 was moving to the launching pad.

On July 16, 1969, three men waited for liftoff in Apollo 11; two of them would walk on the moon.

At 9:32 A.M., the word was "Go!"

CHAPTER

"TRANQUILLITY BASE HERE": MAN ON THE MOON

Shortly after 4:17 P.M. Eastern Daylight Time on July 20, 1969, the Apollo 11 Lunar Module *Eagle* dropped gently to the surface of the moon to place a new name on the lunar landscape—Tranquillity Base. The technology that carried men to the moon also made it possible for the whole world to watch on television as Neil Armstrong and Edwin Aldrin took their first steps on the moon a few hours later.

About 500 million TV viewers back on earth watched as men walked on the moon. Many people found the event hard to believe, even while seeing it take place. The harsh lighting, the sharp, black shadows, the slow movements in the moon's weak gravity, and the almost ghostly appearance of the spacesuited astronauts gave many viewers the feeling that they were watching an old and poorly made science-fiction movie instead of the actual exploration of the moon.

If the first moonwalk seemed unreal, it was partly because the audience was viewing it with perceptions and reflexes conditioned by a lifetime spent in the heavy gravity of earth. The floating strides of the astronauts and the slow collapse of the spurts of dust that they kicked up seemed artificial and contrived. I remember that part of my own mind kept comparing the two spacesuited figures to underwater divers, whose slow motions are familiar and therefore "normal." The Apollo 11 astronauts made a quick adjustment to lunar gravity, but it took a few more missions before the floating motions seemed natural to the rest of us.

For the participants in the Apollo Program, the landing was the climax which justified nearly ten years of continual effort. For the American TV audience, there was the thrill of being present at a major historical event, as well as a strong sense of national pride in an American success. But the moon landing also represented a larger success for all mankind. Apollo seemed inconsistent with narrow national boundaries. The American flag that the astronauts placed on the moon was raised as a symbol of achievement, not as a mark of conquest or a claim for sovereignty.

Science quickly followed the flag to the moon. Armstrong and Aldrin began to gather new information from the moment that they stepped onto the lunar surface. They tested the footing and began to walk, first with cautious hops and then with confident floating strides. Within minutes they were loping easily. "Isn't this fun?" remarked Armstrong.

They described the surface on which they stood: a flat, colorless plain of powdery soil, dotted with small craters and rock fragments. It was, Aldrin said, "a magnificent desolation." There were no traces of the action of wind or water. There was no sign of life anywhere.

Everything they saw was new. They took color photographs of lunar features that men had never seen before: small craters, large rocks sticking up through the loose soil, strange unexplained ridges and furrows on the surface, and gleaming patches of glass splashed into craters and onto the surfaces of rocks.

They collected the samples of rock and soil that a thousand scientists back on earth were waiting for. It turned out that Apollo 11 landed about 6 kilometers away from its intended site in the Sea of Tranquillity; no one was sure exactly where the astronauts had been until after they returned to earth. But it didn't seem all that important. On the flat surface of the Sea of Tranquillity, one point is pretty much like another. What *was* important was that moon rocks were being collected for the first time. Pinpoint landings could wait for later missions.

Part of the astronauts' two-hour moonwalk was devoted to setting up instruments which would collect data about the moon long after they had left it. One of Aldrin's duties was to carry two instruments about 10 meters away from the lunar module, out of range of the blast from the ascent rocket engine, and set them up.

One instrument, a seismometer, was the offspring of instruments long used on earth to detect and analyze vibrations caused by earthquakes. The seismometer that Apollo 11 carried was sensitive enough to record any moonquakes that occurred, and scientists hoped that it would tell them much about the nature of the lunar interior. When Aldrin turned on the instrument, its radio signals were detected immediately by receivers on earth, transmitted through the world-wide network of NASA communications, and delivered to the Principal Investigators at the Manned Spacecraft Center in Houston, Texas. The seismometer was sensitive enough to detect a small tremor at a distance of a thousand kilometers; during the moonwalk it faithfully recorded the small vibrations that the astronauts' footsteps made on the moon.

Another instrument, the Laser Ranging Retro-Reflector, was a precise mirror which would reflect an intense beam of laser light shot up to the moon through a telescope on the earth's surface. By bouncing the laser beam off the reflector and back to earth, astronomers could measure the distance between the earth and moon with an accuracy of about 15 centimeters (less than one foot) in almost 400,000 kilometers (240,000 miles). This is about 5,000 times more accurately than surveyors on earth can measure the 5,000-kilometer (3,000-mile) distance between New York and San Francisco.

By making repeated accurate measurements of the distance between the earth and moon during the coming years, scientists will be able to follow the fine details of the moon's movements as it swings around the earth. From the nearly undetectable dips and wiggles in its motions, the nature of the lunar interior can be more completely understood. It may even become possible to use the data in reverse, using laser reflectors on the moon to detect and measure the slow movements of continents and ocean basins on the earth.

Another instrument placed on the lunar surface looked outward to the stars. Apollo 11 had transformed the hopes of physicists and astronomers into a reality. Now the airless moon was being used as a platform to see the universe more clearly than had been possible through the thick, unsettled atmosphere of earth. The first instrument to take advantage of this new observation platform was probably the simplest mechanism carried to the moon by Apollo 11. It was a strip of aluminum foil mounted on a pole. Aldrin set it up by simply poking the pole into the lunar soil, unrolling the aluminum foil and turning it to face the sun (photo 9). This instrument, known formally as the Solar Wind Composition Experiment and informally as the "windowshade," was designed to trap a sample of the sun itself.

There is no air in space, but there is a "wind." In addition to sending light and heat through the solar system, the sun also sends out streams of single atoms that have been blasted out of its incredibly hot atmosphere and shot into space at speeds of several hundred kilometers a second. This stream of atomic particles, called the solar wind, is a sample of the matter that makes up the sun.

All life on earth depends on the sun, yet we know very

little about it. Only within the last fifty years have we discovered that the sun is a giant nuclear reactor, consuming its hydrogen atoms to form helium and releasing energy at a rate equal to the explosion of a billion hydrogen bombs every second. Even now, while we take the sun for granted as a permanent and unchanging thing, we are slowly learning that the sun, like a human being, has a life history.

The sun was born about 4½ billion years ago in the hot center of the collapsing dust cloud that formed the solar system. It is now in a steady, quiet middle age. In another few billion years it will grow old and die, perhaps quietly cooling so that all the planets freeze, or perhaps giving up its energy in a sudden explosive burst that will turn the earth and the other planets into cinders.

But the death of the sun is of less immediate concern than its present behavior. The sun provides the energy that drives our weather systems and controls our climate. The existence of life on earth depends on the fact that the temperature hovers between the freezing and boiling points of water. A small increase in the sun's energy output could make the earth too hot to inhabit; a small decrease could freeze the oceans and bring back the great ice ages.

Like a doctor examining a patient to find out about his health, scientists need to analyze the sun, to understand the processes that go on inside it, and to determine its state of health, for the future of the sun determines our own future. Before Apollo, scientists had to estimate the composition of the sun indirectly from the nature of its light. The solar wind, composed of actual atoms from the sun, offers information that we could never obtain by indirect observations.

However, none of the solar wind reaches the surface of the earth. Hundreds of kilometers up, the fast-moving particles are blocked by the earth's atmosphere and magnetic fields, and give up their energy to produce the brilliantly colored auroras that we call Northern and Southern Lights. Our atmosphere, which protects us from these streams of atoms, also prevents us from learning what they are.

Therefore, in order to sample the sun, man had to go to the moon. While the astronauts were on the lunar surface, the aluminum "windowshade" trapped and held the atoms of the solar wind that struck it. At the end of their stay, the astronauts rolled up the aluminum foil with its captured atoms and brought it back to earth for analysis.

Even though the solar wind probably contains atoms of nearly all the elements that occur in nature, the Solar Wind Composition Experiment could detect only those elements which were abundant, which did not react chemically with the aluminum foil, and which could be removed from the foil by heating. These limitations meant that only the so-called "noble gas" elements could be analyzed. Thus, helium and neon were observed; whereas argon, krypton, and xenon, if present, could not be detected. Helium, which is the second most abundant element (after hydrogen) in the sun, was also the most abundant element found in the "windowshade"; the scientists estimated that each square centimeter of the aluminum foil had been struck by about 6 million helium atoms and 15,000 neon atoms every second.

The Apollo 11 "windowshade" experiment was an important beginning. Other aluminum sheets, exposed on longer missions, later captured some of the rarer atoms in the solar wind. The lunar rocks and soil provided samples of solar wind that had been trapped thousands, or even millions, of years ago. At last scientists were able to compare the past record of the sun's behavior with its present condition. If we could devise a chronology from the parts of the record, we might know what nuclear reactions had occurred inside the sun in the past, what transformations were going on now, and what might happen to the sun in the future. The pieces of the moon that Apollo 11 brought back were important and exciting, but the pieces of the sun that were carried back in that thin strip of aluminum foil may turn out to be even more important.

EARLY APOLLO: MAKING HASTE SLOWLY

Like a baby's first step, the landing of Apollo 11 was a great event, but it was only the beginning, for the results of Apollo 11 were quickly surpassed by new moon landings (photo 10). With the later missions, we gained more experience with space travel and more confidence in the equipment, and we really began to use the Apollo transportation system to explore the moon. Each new mission was more complex and ambitious than the last, and each landing brought new instruments to the moon.

No one had ever expected that the moon could be studied adequately with a single manned landing. Imagine a Martian astronaut beginning the exploration of the planet earth by landing on the level plains of central Kansas. The samples he collected would provide general information about the planet, and his observations would tell Martian scientists a great deal about what flat farmland was like. But a landing in Kansas would tell nothing about the Rockies, the Appalachians, the Great Lakes, or the Mojave Desert. If this Martian space program stopped with a single landing, their scientists would learn nothing about the mountains, oceans, forests, volcanoes, and glaciers that are so essential to understanding what the planet earth is really like.

Scientists on earth were in the same position after Apollo 11. The samples returned from Mare Tranquillitatis taught us a lot about the lava flows that covered one part of the moon. But there were other regions of the moon that still needed to be investigated, and later missions focused on other maria, the rounded Apennine Mountains near Hadley Rille, and the rugged highlands near the crater Descartes.

Only four months after the first footsteps on the moon, Apollo 12 made a pinpoint landing in Oceanus Procellarum. Astronauts Charles Conrad and Alan Bean landed the

Lunar Module *Intrepid* less than two hundred meters from the Surveyor 3 spacecraft that had been sitting silently on the moon since its landing there two and a half years before. The astronauts remained on the moon for several hours, taking two long walks and collecting samples according to a prearranged plan. Their samples came from a variety of lava flows which were both younger and chemically different from the samples brought back by Apollo 11.

Instead of just setting out a few separate instruments, Conrad and Bean deployed a complete scientific station, including a second seismometer to detect moonquakes and several different instruments designed to measure the moon's weak magnetic field, to search for traces of gas molecules near the lunar surface, and to record the bombardment of the moon by matter and energy from space. The whole station had been planned and constructed so efficiently that it used less power than a 100-watt light bulb to make all the measurements and transmit them back to earth.

The Apollo 12 astronauts were instructed not to go farther than a few hundred meters from their Lunar Module, so that they could get back to it quickly in case any problems developed with the cooling and oxygen supply systems in their spacesuits. But Apollo 12's pinpoint landing made it possible for the astronauts to reach the nearby Surveyor 3 spacecraft in order to salvage its TV camera (which had a glass lens) and some pieces of metal from the spacecraft itself, thus giving scientists an opportunity to discover exactly what had happened to pieces of glass and metal left on the moon for a known period of time.

A thorough examination of these objects produced no evidence that even the tiniest meteorite had hit the spacecraft in two and a half years. On the basis of these results, scientists felt more confident that meteorites would not endanger astronauts, even if they were to remain on the lunar surface for several days. Biologists found no trace of microscopic lunar life on the pieces of the spacecraft. Finally, the data from Surveyor 3 were used to calculate how fast the lunar surface was being worn down by particles bombarding it from outer space. Erosion on the moon goes on unbelievably slowly; it takes about 50 million years to wear down the lunar surface by only a millimeter. The moonscapes that we now see have existed unchanged for billions of years. Compared to the earth, where wind and water can completely change the surface in a few years, the face of the moon seems permament and practically unchanging.

An even larger scientific program was planned for Apollo 13. Unfortunately, the mission turned into a near disaster on the third day after launch when an oxygen tank exploded, crippling the spacecraft's power supply and making the lunar landing impossible. All at once, the whole world was sharply reminded that, even after two successful lunar landings, space travel was still dangerous and unpredictable. At the time of the accident, Apollo 13 was over 320,000 kilometers (200,000 miles) away from earth, more than three-quarters of the way to the moon, and the quickest way back to earth was to swing around the moon first. The safe return of the Apollo 13 crew proved that men and equipment could handle unexpected crises; the crew survived by using the Lunar Module as a "lifeboat," living on the power and oxygen that they would have used during their stay on the moon. Although Apollo 13 never landed on the moon, the crew obtained some excellent photographs of its far side as they swung around it to begin their return trip—pictures taken by disciplined men who were not sure at the time that they would return safely to earth to show them.

The ambitious exploration planned for Apollo 13 was carried out instead by Apollo 14 in January, 1971. The astronauts Alan Shepard and Edgar Mitchell landed their lunar module *Antares* not far from the Apollo 12 site, near a large crater called Fra Mauro. They made a long traverse on foot through rolling hills and boulder fields to the rim of a small crater named Cone. They set up another scientific station and thus helped to build up a network of recording instruments across the moon. For the first time on the moon, they set out sensitive earphones, then set off small explosions on the ground, while they listened for echoes that would tell them about the layers of rock and soil beneath them. They measured the moon's magnetic field on the surface and discovered that it was much stronger than anyone had thought possible. Finally, they brought back pieces of the unusual Fra Mauro Formation, strange rocks made up of light and dark materials mingled together (photo 11).

The return of Apollo 14 ended the group of "Early Apollo" missions. These missions had established that the Apollo equipment could take men to the moon, support them while they were there, and bring them back safely. The landings had also shown that men could walk and work on the moon. Astronauts had described and photographed the lunar surface, picked up loose rocks, broken pieces off boulders, driven sampling tubes into the soil, and set up scientific instruments. In short, the Early Apollo missions had demonstrated that men on the moon could do everything necessary for scientific exploration (*Figure C*).

Back on earth, the lunar samples were providing scientists with a steady stream of information about the great age of the moon and the huge impacts and great floods of lava that had figured in its history. The formerly sketchy outlines of the origin and history of the moon were becoming clearer, and the all-out study of the moon was about to begin.

Figure C / Working Clothes for the Moon. This diagram shows the variety of equipment carried by spacesuited astronauts for scientific exploration of the lunar surface: sample-collecting tools, bags and core tubes for storing samples, cameras, a watch, checklists, and a Portable Life Support System (PLSS) which supplies oxygen and cools the spacesuit. After collection, samples are carried in a backpack next to the PLSS until the astronaut returns to the Lunar Module. (Based on NASA photograph 72–H–180)

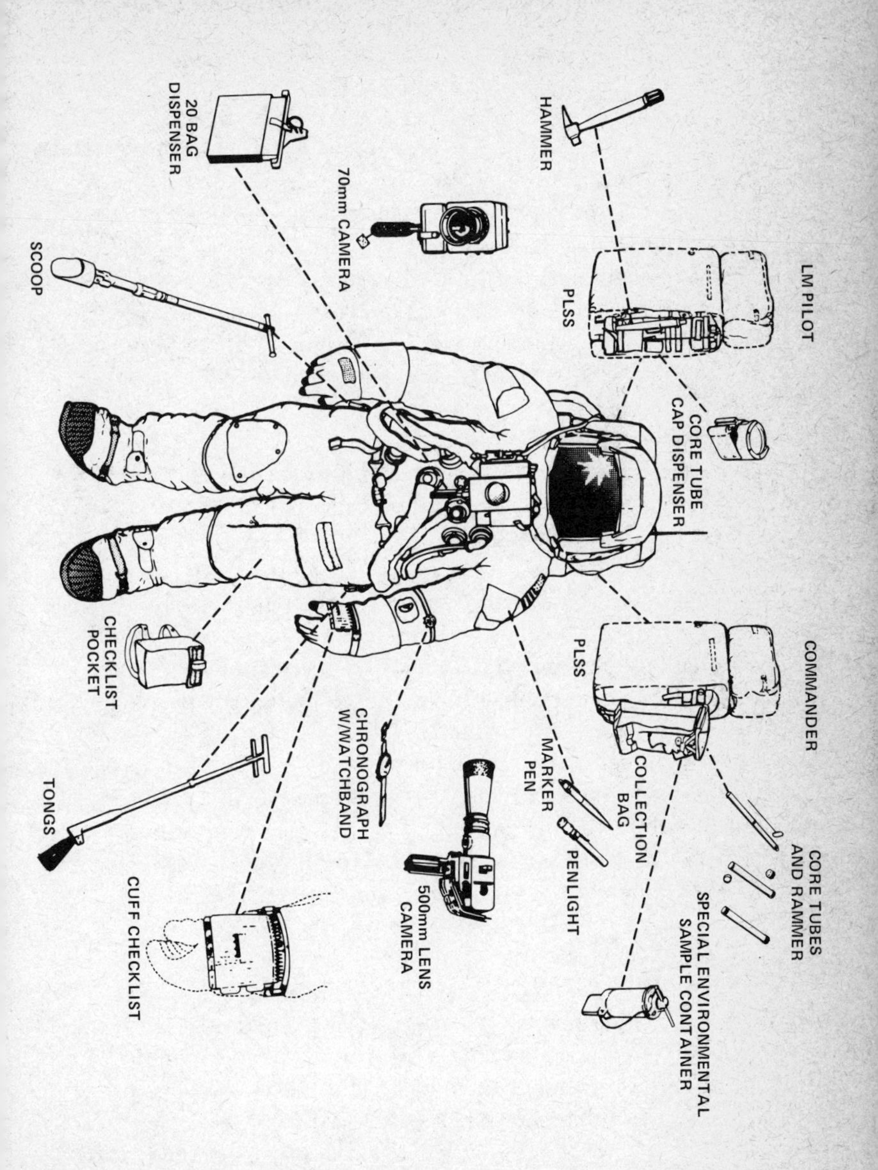
LM PILOT
PLSS
HAMMER
CORE TUBE CAP DISPENSER
70mm CAMERA
20 BAG DISPENSER
SCOOP
COMMANDER
PLSS
CORE TUBES AND RAMMER
SPECIAL ENVIRONMENTAL SAMPLE CONTAINER
COLLECTION BAG
PENLIGHT
MARKER PEN
500mm LENS CAMERA
CHRONOGRAPH W/WATCHBAND
CHECKLIST POCKET
TONGS
CUFF CHECKLIST

LATE APOLLO: PAYLOADS AND PAYOFFS

Apollo 15 looked just like any other Apollo launch as it thundered into the clear sky above Cape Kennedy, almost two years to the day after men had first stepped onto the moon. But Apollo 15 was the first of a series of missions that would focus entirely on the prolonged and intensive scientific exploration of the moon. For these flights, the original Apollo system had been modified to carry more equipment to the moon and to bring back heavier loads of lunar samples. The new mission, planned with the experience gained from two years of successful lunar exploration, would carry astronauts to more rugged and difficult landing sites selected by eager scientists: Hadley Rille, the Descartes highlands, and the Littrow Valley.

The explorations carried out by Apollo 15, 16, and 17 were to make the accomplishments of Apollo 11 seem cautious and tentative. These later missions provided for the astronauts to remain on the moon for days instead of hours. Instead of walking only a few hundred meters, they rode for many kilometers in a Lunar Rover. Instead of hurriedly picking up rocks at random, they made educated observations about the lunar surface and used their training in geology to select and document the samples they collected. Mission after mission brought back hundreds of pounds of many different rocks and soils from new maria, rilles, and highlands.

A whole new generation of instruments went to the moon on these later flights. Each mission carried over 20 different scientific experiments. Additional seismometers and laser reflectors were placed on the surface, gradually expanding a moon-wide instrument network that was already beginning to probe the lunar interior. Electric drills cut out cores two meters long from the lunar soil so that scientists could read the hundreds of millions of years of lunar history preserved in the thin layers. Also among these instruments was an important device called the Heat Flow Experiment, which scientists hoped to use to measure the temperature of the moon and to answer the long-debated question of whether the inside of the moon is hot or cold.

Determining the internal temperature of the moon was one of the chief goals of the Apollo Program. Because the amount of internal heat determines what a planet is like, this information provides scientists with an important clue to its formation and evolution. A body with little internal heat will remain "frozen," whereas a hot planet like the earth will have the energy to undergo great changes, producing earthquakes and volcanic eruptions, building up great mountain ranges, and moving continents and ocean basins around on giant plates.

There are several ways in which heat can be generated inside a planet. Some heat can be produced when the planet is formed by small meteorites and planetoids that come together to build up a larger object. As the planet grows larger, its gravitational force increases. It attracts other bodies more strongly, causing them to strike its surface at higher and higher speeds. These high-speed collisions release a great deal of energy in the form of heat, which may remain in the outer part of the planet for billions of years.

Even after a planet has stopped growing, it can still generate internal heat. Heavy material, such as molten iron, may move from the outer part of a planet down toward the center. As this material moves from a high level in the planet to a lower one, it releases gravitational energy, much as water gives up energy (to turn a mill wheel or run a generator) as it flows from a higher elevation to a lower one. Inside a planet, this energy is released as heat, and the formation of a core of metallic iron can release enough heat to melt a large part of the rest of the planet.

These two sources of heat can exist only during the early

history of a planet. If they were the only sources of planetary heat, then all the planets would have been cooling steadily for about the last 4½ billion years. However, there is another source of internal heat that can operate throughout the whole lifetime of a planet. The discovery of radioactivity about a century ago forced scientists to consider the significant amounts of heat that are released by the slow and steady decay of the tiny amounts of radioactive elements, such as uranium and thorium, that occur in all natural rocks.

Men have known for centuries that the earth is hot inside. The great floods of molten lava that repeatedly pour out of volcanoes have always been impressive proof of our planet's internal heat. More recently, when deep mines were dug and oil wells drilled, men found that the deeper they went the hotter it became. In most places the temperature rises 1° to 3° C. for each hundred meters of depth. The world's deepest mines (in South Africa) extend nearly 4 kilometers down; the temperature at the bottom (about 55° C., or 130° F.) makes extensive air conditioning necessary.

It is a basic law of nature that heat flows from hot objects to colder ones. Thus, heat is continually flowing outward from the interior of the earth to the surface, where it can be detected and measured by sensitive instruments.

The amount of heat lost by the earth is very small—only a few millionths of a watt per square centimeter of surface. It would take the earth's heat flow, unaided, about five years to melt an ice cube placed on the surface. This tiny amount of heat, then, is totally lost in the larger amounts of heat supplied to the surface by sunlight and atmospheric circulation.

But even tiny amounts of heat are important. Added up over millions of square kilometers of surface area and billions of years of time, this internal heat generates the volcanoes and earthquakes that have played such a major role in the earth's history.

Thus, the experimental measurement of the heat flow on the moon's surface also provides a way of probing the lunar interior. Scientists already knew that the moon had been hot about 3½ billion years ago, when some part of the moon melted and lava flows had poured onto the maria. What was not known was how deep in the moon this melting had occurred and whether the moon might still be partly molten inside. The Apollo 15 Heat Flow Experiment attempted to answer these questions. At the same time the data would provide important chemical information: from the amount of heat observed, scientists would be able to calculate how much radioactive material the interior of the moon contained.

Heat flow measurements on the moon were made with the same techniques that had been used for years to determine terrestrial heat flow in deep mines and in the mud at the bottom of the sea. The Apollo 15 astronauts drilled two holes 1 to 2 meters deep and about 2 centimeters in diameter into the lunar soil. Into each hole they lowered a long probe with sensitive electrical thermometers placed along its entire length (*Figure D*). Data from the thermometers revealed that the temperature was 1° C. higher at the bottom of the hole than at the top.

Before the Apollo landings, theoretical estimates of what the lunar heat flow would be had ranged between 1 and 5 microwatts per square centimeter (a microwatt is a millionth of a watt). The range of predicted values resulted from the great variety of chemical compositions proposed for the interior of the moon. The higher value was close to the average heat flow of the earth itself, and most scientists, especially those who felt that the moon was a "cold" body, expected much lower values of 1 microwatt or less.

The data from the thermometers in the heat flow instrument came as a great surprise to everyone. The moon was much hotter inside than most scientists expected. The temperatures inside the lunar soil increased rapidly downward, about 1.8° C. for each meter of depth.* These measurements, combined with information about the thermal properties of the lunar soil, made it possible to calculate the heat flow coming out of the moon. The value obtained was 3.3 microwatts per square centimeter, about half that of the

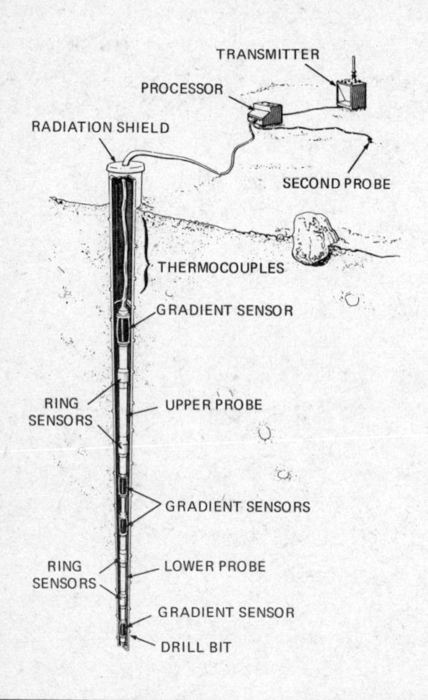

Figure D / Taking the Moon's Temperature. The Apollo 15 Heat Flow Experiment consists of two probes which are lowered into holes drilled into the lunar soil. One probe is shown in cross section; it is about a meter long and contains a variety of sensors (detectors) to measure gradients (temperature differences) in the hole.

* The loose, porous lunar soil is a very good insulator, that is, it conducts heat very poorly. The heat escaping from the inside of the moon thus takes much longer to travel through the lunar soil than it would take to pass through an equally thick layer of solid rock. As a result the temperature increases more rapidly below the surface

earth. That is a large amount of heat to be coming out of a body that has only one-eightieth the volume of the earth.

Naturally, this discovery stimulated a flood of new questions. How could the moon be so hot inside when there was no evidence for recent volcanic eruptions? How could the experimental evidence for a hot lunar interior be reconciled with the evidence for a cold and rigid interior that had been supplied by the discovery of mascons? One compromise model involved the idea of a "warm" moon that would satisfy both lines of evidence. Such a moon would be hot enough inside to generate the observed heat flow, but not so hot as to cause the moon's crust to lose its strength. Another suggested explanation was that the heat-producing radioactive elements had been somehow concentrated in the outer layer of the moon while the inside was still cold. Proponents of this idea ran into problems when they attempted to explain how such an extreme chemical separation could have occurred. More measurements of lunar heat flow and more analyses of the radioactive elements in the returned lunar rocks seemed to be needed to obtain a clear explanation.*

---

in the lunar soil (1.8° C. per meter) than in solid terrestrial rocks (about 0.03° C. per meter). The rapid temperature rise must occur only within the lunar soil layer, for if the temperature continued to rise that rapidly farther down, the entire moon would be molten a few kilometers below the surface. Beneath the thin surface layer of lunar soil, there is solid rock in which the temperature increases much more slowly with depth, so that, even deep inside the moon, temperatures are probably not much higher than about 1,000° C.

* More recent data from the heat flow instruments on the moon indicate that the moon is not quite as hot as we first thought. At the Seventh Lunar Science Conference in March, 1976, a heat flow value of about 2 microwatts was reported; this lower value is more consistent with models for a "warm" moon. The higher values originally reported (about 3.3 microwatts) were the result of imperfect knowledge about the mechanical and thermal properties of the lunar soil; these properties had been altered as the astronauts walked on the soil and inserted the heat flow experiment itself. Now that the instruments have been operating continuously for almost five years, scien-

Other instruments carried by Apollo 15 (and subsequent missions) took advantage of the moon's position as a platform for viewing the rest of the universe. Small telescopic cameras placed on the moon "saw" by a light to which our own eyes are blind; they detected weak ultraviolet rays that cannot penetrate the earth's atmosphere. The pictures quite literally showed us the stars and the earth itself in a new light (photos 12 and 13).

Apollo 15 made a perfect landing in a rugged region selected for its scientific importance. The landing site was at the edge of Mare Imbrium, between the high rounded summits of the Apennine Mountains and the deep, winding canyon of Hadley Rille. With the lunar roving vehicle, astronauts Jim Irwin and David Scott made an extensive geological study which carried them along the edge of the Rille and out across the front of the ancient mountains that tower above it.

The astronauts observed fine details that no Surveyor TV camera could have detected. They discovered and photographed the fine lines that slant across the surrounding mountains. At first, geologists thought that the lines might be different layers of rock that had been deposited horizontally and then suddenly raised and tilted by the catastrophic impact that created the Imbrium Basin. Further study of the pictures by geologists on earth suggested that the lines might be caused by the effect of the harsh sunlight shining at a low angle across the lunar surface.

Irwin and Scott were the first astronauts to see solid lunar bedrock still in its original position in the walls of Hadley Rille. They photographed a sequence of horizontal layers of basalt lava flows that had not yet been broken up by meteorites or buried by the lunar soil. Here was more evi-

---

tists have obtained enough data to correct for this initial disturbance of the soil and to measure both the heat flow and the properties of the lunar soil more accurately.

dence that Hadley Rille might indeed be a huge channel through which molten lava had poured out onto Mare Imbrium.

Despite the number of new instruments carried on the later Apollo missions, the astronauts still had time to collect an increasing number and variety of lunar samples for the continually eager scientists back on earth. The Apollo 15 astronauts broke off pieces of the many different kinds of boulders that lay on the surface. They also collected samples of lunar soil at various points along their route. They returned to earth with 77 kilograms of lunar material and more than 350 separate samples. Many of the rocks were lava flows of a wide variety of textures and chemical compositions. The Apollo 15 mission also brought back the unusual crystalline Genesis Rock (sample 15415) which turned out to be 4 billion years old, as well as a fragile piece of lunar material, dubbed the Green Clod (sample 15426), which was composed of tiny spherules of green glass—the puzzling and beautiful product of some ancient impact or volcanic eruption.

Apollo 16 was the first mission to explore the lunar highlands, in an area near the crater Descartes. The astronauts, John Young and Charles Duke, made two long excursions in their Lunar Rover. First they drove about 5 kilometers south from the landing site and up onto a medium-size hill called Stone Mountain. Their next trip took them 5 kilometers north, to visit North Ray Crater and to collect pieces of lunar bedrock thrown out by the force of the ancient impact which had formed the crater (photo 14).

The Apollo 16 astronauts were instructed to sample a unit of rocks named the Cayley Formation. Geologists had expected that these rocks would prove to be very ancient lava flows that had been erupted in the highlands even before the maria had formed. They could not have been more wrong. The well-trained astronauts on the surface recognized immediately that the rocks were not lavas. The samples they described and collected were breccias, rocks made up of the crushed and shattered pieces of other rocks. No traces of lava flows were found at any of the Apollo 16 sampling sites. All the samples returned were melted or pulverized pieces of the highland rocks themselves.

The Cayley Formation, which provided the Apollo 16 samples, seems to be not lava flows, but a huge accumulation of debris from a major impact on the moon. Some geologists suggested that the large basin of Mare Orientale was the original source of these highland rocks. The catastrophe which formed this 900-kilometer-diameter basin could easily have provided the energy to hurl a blanket of shattered and melted rock more than a quarter of the way around the moon to the Descartes site.

The Apollo 16 mission thus caused a major revision in our picture of lunar history. What had been thought to be an episode of ancient volcanism was now recognized as one of the widespread effects of a catastrophic impact. This unexpected discovery taught us something even more important: no matter how carefully telescopic mapping is done, the actual geology and history of another planet can be determined only by landing and sampling the rocks. On the Apollo 15 mission, the geological predictions had been largely confirmed by the rocks brought back from Hadley Rille. On the Apollo 16 mission, the results were just the reverse, and the mission taught us that some caution will be needed in the future as we try to decipher the history of Mercury or Mars with only photographs of their surfaces to work with.

The Apollo 17 landing in the Littrow Valley at the edge of Mare Serenitatis was an attempt to find both old and young lunar rocks in one place. Serenitatis is an ancient basin, older than Mare Imbrium, and the mountains around it might contain rocks formed during the earliest stages of lunar history. Yet, careful study of Apollo 15 orbital photographs of the Littrow Valley also showed what looked like deposits of young volcanic rocks. The youngest lavas so far returned from the moon had been about 3.3 billion years old. Finding even younger lavas would be very significant, because it would revise our estimate of how long the moon had remained molten inside,

The Apollo 17 astronauts, Eugene Cernan and Harrison ("Jack") Schmitt, landed in the Littrow Valley, a small depression almost totally surrounded by high, rounded hills. Schmitt was one of the new scientists–astronauts, a geologist with a Ph.D. who had given up the study of the ancient rocks of Norway for the chance to find even older rocks on the moon. He and Cernan carried out a thorough and exacting geological traverse, crisscrossing the Littrow Valley in their Lunar Rover, deploying instruments, sampling lavas from the floor of the valley, and collecting pieces of the huge boulders perched on the surrounding hills (photo 15).

The high point of the Apollo 17 mission was the discovery of the spectacular Orange Soil at one of the small craters on the valley floor. Until then, all the astronauts had seen the moon as a place of subdued, almost colorless grays and browns, and the discovery of anything bright orange was totally unexpected. The soil was made up of beautiful, clear glass droplets (photo 16) that seemed so fresh they might have formed yesterday. For awhile, excited scientists thought that the Orange Soil was proof of recent volcanic activity on the moon and speculated about how the glass spherules might have formed in flaming fountains of young lava.

There was surprise and some disappointment when the Apollo 17 samples were studied back on earth. No young rocks were found. The lavas from the valley floor were 3.7 billion years old—the same age as the first lunar rocks returned by Apollo 11. Even the Orange Soil, which seemed so fresh, was 3.7 billion years old. Nor were the rocks collected from the hills any older than the oldest rocks found by other missions. Like earlier missions, Apollo 17 brought back some rocks with measured ages of 4.0–4.3 billion years, but the interpretation of these numbers is still uncertain. In a way, the Apollo missions had come full circle, with the youngest rocks returned by the last mission turning out to be just as old as the rocks brought back by the first.

After the Apollo 17 Command Module *America* splashed safely into the Pacific Ocean on December 19, 1972, the Apollo Program ended. But the study of the moon was just beginning. The 382 kilograms of lunar rock and soil returned by six successful Apollo missions became the center of an increasing scientific effort. Many of the devices left on the moon are still transmitting useful data back to earth. One of the results of this program has been to domesticate the moon, turning it into a huge instrumented satellite that is being used, just like smaller satellites built and launched from earth, to learn more about the universe.

SCIENCE FROM ORBIT

A great deal of the lunar research in the later Apollo missions was carried out by astronauts who never set foot on the moon. Many new scientific instruments, which had not been used for the early Apollo missions, were located in the Apollo Command Module, the spaceship that carried one astronaut in orbit around the moon while the other two worked on the surface below him.

Although the Orbital Science Experiments never appeared on living-room television screens during moonwalks, these activities were just as important as the surface work. Even with unlimited Apollo landings, man could not study the whole surface of the moon on foot. The Command Module carried a battery of instruments to photograph, measure, and analyze the large areas of the moon that the Apollo Program would never reach (photo 17).

As the Command Module orbited the moon 125 kilometers above the surface, it photographed and analyzed the lunar surface directly beneath it. Because the orbits of most of the Apollo missions were close to the lunar equator, it was not possible to study the total surface, but nearly a quarter of the moon was covered. Apollo 15, 16, and 17 performed the impressive task of photographing and analyzing an area of the moon as large as the United States and Mexico combined. More important, half of this region lay on the far side of the moon, where scientific information is hard to obtain. The orbital experiments would be the first scientific examination of the moon's far side since the region had been photographed by the Lunar Orbiters several years before. They would also be man's last look at this region for some time.

Among the orbital instruments were several film cameras that were technically far superior to the ones used by the Lunar Orbiters. Placed next to the cameras was a laser altimeter, which shot an intense beam of laser light down to the lunar surface. This instrument measured the heights of mountains and the depths of valleys to an accuracy of within a meter from an altitude of 125 kilometers. The measurements showed just how rugged the moon really is. Some highland areas on the far side of the moon rise almost 4 kilometers above the average surface level. There is a drop of about 8 kilometers from the highest lunar mountains to the floor of the maria. Furthermore, some of the lunar maria on the earth-facing side of the moon were found to lie about 4 kilometers below the average level of the lunar surface.

One of the most ambitious undertakings of the orbital science program was a chemical analysis of the moon. As the Command Module passed over the moon's surface, its instruments also analyzed the earlier Apollo landing sites, whose exact chemistry was known from the returned samples. The combination of analyses from lunar orbit with the results from the Apollo samples made it possible for scientists to extend the geological results of the Apollo landings to nearly a quarter of the moon's surface.

To make the chemical analyses, the instruments in the Command Module used the X-rays that are emitted by the sun along with its light and heat. The sun's X-rays are very weak, and they are easily blocked out by the earth's atmosphere, but they continually reach the surface of the airless moon. These X-rays are no danger to an astronaut, for they are too weak to penetrate deeply into any solid material like a spacesuit. Traveling only a few hundredths of a millimeter into the lunar soil before stopping, these solar X-rays excite the atoms in the soil and cause them to give off weak X-rays of their own. Each different chemical element in the soil emits its own characteristic X-ray, and these secondary X-rays are radiated back into space. High above the lunar surface, special instruments in the Command Module detected the different X-rays and sorted them out, thus determining the amount of each different element present in the lunar soil.

The same principle is used routinely on earth in X-ray machines that make chemical analyses of rocks and minerals. But the Apollo experiment was using the sun as an X-ray tube and the entire moon as a target. In one orbit of the moon, these X-ray detectors obtained chemical information from a larger area of the moon than could be obtained from centuries of Apollo landings and surface explorations.

The most important result of these chemical analyses was the discovery that there were definite chemical differences between the dark maria and the lighter highland areas. The origin of the light and dark colors on the moon, which had been a topic of speculation since before the ancient Greeks, was finally settled. The maria and highlands are different colors because they are made up of different rocks. The light-colored highland rocks are rich in calcium and aluminum, whereas the maria are made up of darker rocks that are rich in such elements as magnesium, iron, and titanium. The highlands and maria are fundamentally different in their chemistry and origin, and these chemical differences must have existed more than three billion years ago when the maria formed.

The fact that these chemical differences could still be detected in the lunar soil after three billion years told us something more about the moon. The orbital detectors were analyzing only the uppermost part of the lunar soil, a layer less than a tenth of a millimeter thick. But comparison of the orbital results with the analyses of returned lunar samples showed that this very thin surface layer faithfully preserves most of the chemical characteristics of the bedrock beneath it. If chemical differences can still be detected on the moon's surface, then however the lunar soil was formed, even its uppermost part is not well mixed. Had the lunar soil been evenly mixed and distributed over the entire surface of the moon, all the chemical analyses from orbit would have given the same average results, regardless of whether the highlands or the maria were being analyzed.

These orbital analyses thus confirmed the conclusions of the scientists who had studied the pieces of the Surveyor 3 spacecraft returned by Apollo 12. Erosion and change take place so slowly on the moon that even three billion years is not long enough to move much soil from one part of the moon to another.

While the X-ray instruments in the Command Module were analyzing the lunar surface, other detectors were scanning the moon for traces of radioactivity. Radioactive elements like potassium, thorium, and uranium are crucial to our understanding of the moon because they are important sources of heat. These elements are very rare in lunar rocks—so rare that they are generally measured in parts per million instead of in percents. But on the moon, as on the earth, their accumulated radioactive heat, built up for millions of years, is capable of melting the interior and producing the floods of lava that cover the maria.

The detectors revealed that the distribution of surface radioactivity on the moon was not uniform. What they found were concentrations, or "hot spots," of radioactive elements; one large concentration appeared around the margin of Mare Imbrium. It seemed that the rocks around Mare Imbrium were slightly more radioactive than rocks in other locations. If the radioactive rocks had been blasted out of Mare Imbrium by the impact 4 billion years ago, then it was likely that the interior of the moon in this region contained more radioactive elements than it did elsewhere.

Scientists wondered whether there was a connection between this higher radioactivity and the lava flows that had filled Mare Imbrium and spread over much of the near side of the moon. If the radioactive heat under Mare Imbrium had provided the energy needed to generate the lavas, then why, they asked, were the lavas formed 300 million years *after* the Imbrium Basin itself? While these questions were being debated, other "hot spots" were detected on the far side of the moon, but these were found in areas containing practically no lavas. These other concentrations of radioactivity are even harder to explain. Perhaps they were never hot enough to produce lavas and are now cooling slowly. Or perhaps they are slowly heating up to become the sites of future volcanic eruptions.

The orbital instruments were actually searching for several different kinds of radioactivity. One set of detectors had found "hot spots" of uranium, thorium, and potassium which served as records of lunar events that had happened billions of years ago. Another instrument was looking for accumulations of the radioactive gas radon, which might report events only a few weeks old.

The search for radon was really an attempt to detect recent volcanic activity on the moon. Radon is produced by the radioactive decay of uranium in the lunar interior. Because it is a gas, radon migrates quickly to the surface through cracks and crevices in the rock. Because it decays into other elements quickly, any radon detected on the moon's surface must have come up from the lunar interior in less than a week or two. The detection of radon would pinpoint areas where gases were leaking rapidly out of the moon; such places might be sites of recent volcanic activity.

The instruments found evidence that radon was present on the surface of the moon. Furthermore, the distribution of radon, like the other radioactive elements, was not uniform. Concentrations of the gas were detected near the crater Aristarchus and around the edges of many of the circular maria. But this discovery did not prove that any major volcanic activity was taking place. The orbital instruments were so sensitive that they could detect an amount of radioactive radon equivalent to only a few thousand atoms per square centimeter of lunar surface. Such a small amount could be provided by slow leakage out of the moon without requiring eruptions of larger amounts of other gases.

Still, the discovery that at least some gaseous material is coming out of the moon is significant. As the Apollo Program ended, the moon seemed cold and quiet, but perhaps not completely dead.

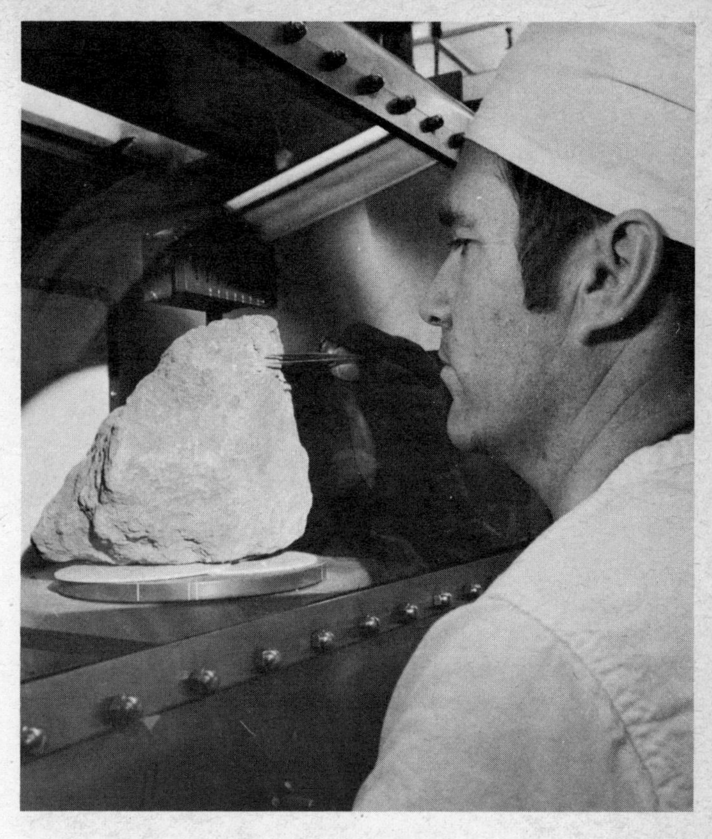

1 / *Man Meets Moon Rock.* A scientist at NASA's Lunar Receiving Laboratory in Houston, Texas, examines a large rock returned from the moon. The rock is sealed in an airtight cabinet to protect it from the oxygen and water in our atmosphere. *(NASA photograph 71–H–472)*

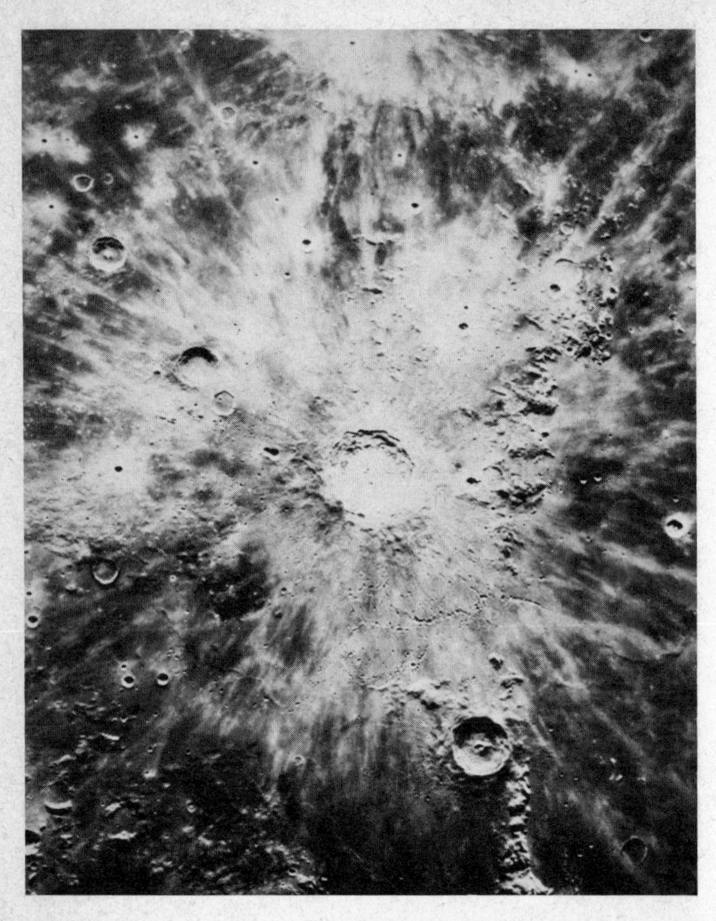

2 / *A Giant Impact Scar on the Moon?* Seen through a large telescope on the earth, the crater Copernicus and its spectacular system of bright rays stand out against the lunar surface. The crater is 90 kilometers (56 miles) across, and the bright rays that radiate from it were probably produced by material thrown out in all directions when the crater was formed. *(Mount Wilson Observatory photograph, reproduced by permission of the Hale Observatories)*

3 / *A Giant Lunar Volcano?* The large crater Plato (at upper left), shown in telescopic view from the earth, is 90 kilometers (56 miles) across, identical in size to Copernicus. Plato, however, is filled with the same dark material (basalt) that covers the surface of Mare Imbrium (at lower left), and the crater may be a giant volcano formed when the lavas erupted onto the moon. *(Lick Observatory photograph, reproduced by permission)*

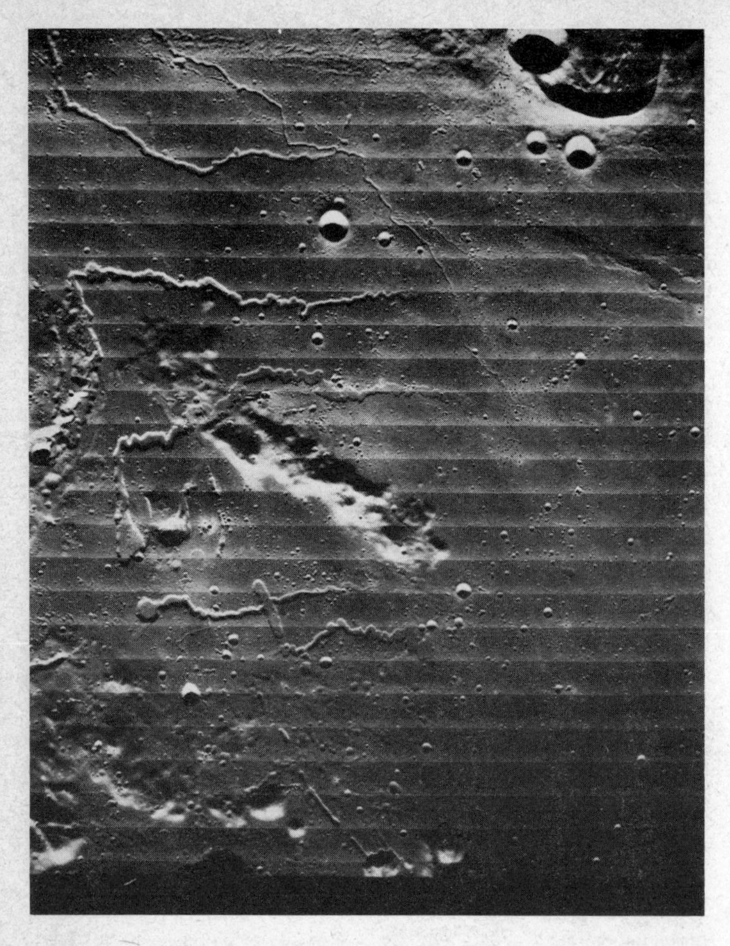

4 / *Rivers or Lava Channels on the Moon?* A group of winding canyons *(sinuous rilles)*, photographed by Lunar Orbiter 5, cuts across the surface of Oceanus Procellarum, near the crater Prinz. The long dimension of the photograph covers about 110 kilometers (70 miles) of the moon. *(NASA Lunar Orbiter photograph V–191–M)*

5 / *A Map of Moon Rocks* (facing page). A geologic map, prepared from observations through an earth-based telescope, shows the different rock formations in the region of the moon near the Apollo 15 landing site. The oldest rocks (Unit 1, medium gray) were present before the

basin of Mare Imbrium was formed, and they now form the Apennine Mountains. Younger rocks (Unit 2, dark gray) are blankets of material thrown out of large craters like Archimedes (upper left), which developed after Mare Imbrium was formed but before it was filled with lava flows. The next youngest rocks (Unit 3, light gray) are the widespread lava flows that cover Mare Imbrium (at left) and cover part of the blanket of material thrown out of Archimedes. The youngest rocks (Unit 4) are small blankets of material thrown out of younger meteorite impact craters.

The Apollo 15 landing site is shown by a triangular arrow. Hadley Rille is the thin dark line nearby, and it and the other rilles in the region are about the same age as the lavas that fill the maria, about 3.3 billion years old.

The map covers an area more than 400 kilometers (250 miles) square. The crater Archimedes (at upper left) is 69 kilometers (43 miles) across. *(The map is taken from the publication* Geologic Map of the Montes Apenninus Region of the Moon, *by R. J. Hackman, U.S. Geological Survey Map I–463, (LAC–41), 1966.)*

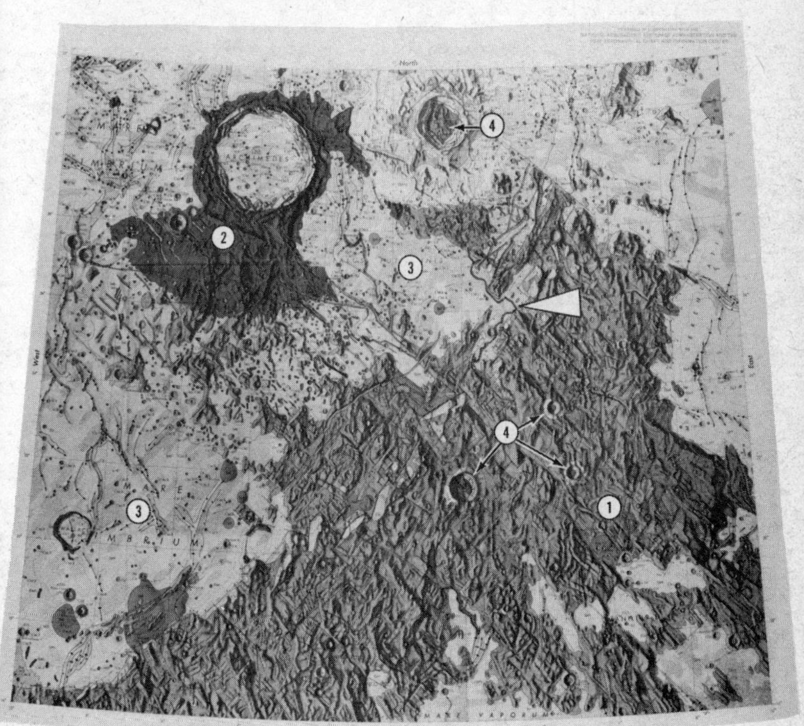

6 / An Orbiter's-Eye View into a Huge Lunar Crater. The interior of the crater Copernicus, 90 kilometers (56 miles) across, is shown in awesome detail in an Orbiter 2 photograph taken from just above the crater rim. The far wall of the crater appears as a series of terraced steps at the top of the picture. This wall is more than 3 kilometers (2 miles) high, or twice the distance from the top to the bottom of the Grand Canyon. *(NASA Lunar Orbiter photograph II–162–H)*

7 / *A Rare "Black Eye" on the Back Side of the Moon.* The crater Tsiolkovsky, named after a Russian pioneer in rocketry, contains one of the rare deposits of dark mare material, probably basalt, on the back side of the moon. *(NASA Lunar Orbiter photograph III–121–M)*

8 / A Field of Volcanoes on the Moon? A Lunar Orbiter panorama shows a large number of low domes and hills on the surface of Oceanus Procellarum, near the large crater Marius, 40 kilometers (25 miles) across, shown in the distance at upper right. These domes, called the Marius Hills, are younger than the lavas that cover the surface of Oceanus Procellarum, and they may have been formed by small eruptions of thick lava that did not spread far over the surface. Similar features are found in many terrestrial volcanic fields. *(NASA Lunar Orbiter photograph II–213–M)*

9 / *Collecting Pieces of the Sun.* Astronaut Edwin Aldrin, standing on the lunar surface, unrolls the "windowshade" experiment, a sheet of aluminum foil. The aluminum sheet, which traps atomic particles of the solar wind, is then brought back to earth for analysis. The Lunar Module, partly in shadow, is in the background. *(NASA photograph AS–11–5872)*

N
W
E
A-15
A-17
A-11
A-12
A-14
A-16
MARE IMBRIUM
MARE SERENITATIS
MARE CRISIUM
MARE TRANQUILLITATIS
MARE FOECUNDITATIS
MARE NECTARIS
FRIGORIS
OCEANUS
PROCELLARUM

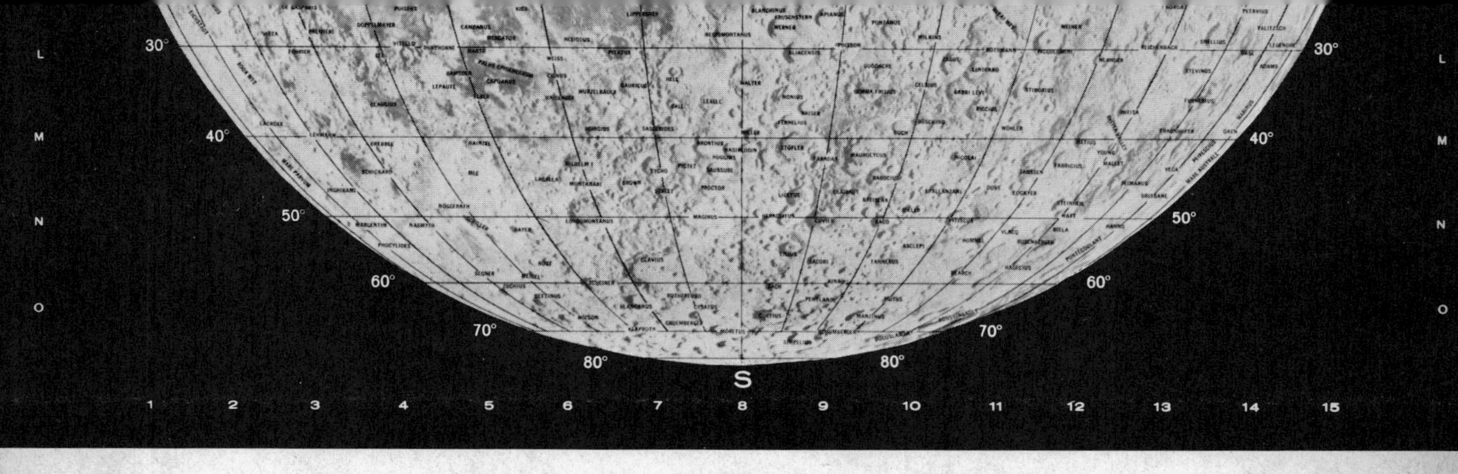

10 / Footsteps on the Moon. The locations of the six Apollo lunar landings are marked on this map, showing the areas studied by Apollo 11, 12, 14, 15, 16, and 17. Apollo 13 returned to Earth without landing. *(NASA photograph 72–H–183)*

11 / A Strange and Mixed-Up Lunar Rock. Looking like a well-mixed marble cake, a ten-foot boulder of breccia from the Fra Mauro Formation waits on the lunar surface to be sampled by the Apollo 14 astronauts. The dark material which makes up the upper part of the boulder is mixed with the white material which makes up most of the lower part. Both the light and dark parts are complex microbreccias which record a very complicated history for the whole rock. *(NASA photograph AS14–68–9448)*

12 and 13 / The Earth in a New Light. A special camera carried on the Apollo 16 mission takes a picture of the earth in ultraviolet light that never penetrates the earth's atmosphere. In an enhanced photograph (left), the earth itself, 12,900 kilometers (8,000 miles) in diameter, occupies the bright central part of the picture. The camera has detected, for the first time, the light from a cloud of hydrogen that surrounds the earth to a distance of about 80,500 kilometers (50,000 miles). For comparison, the familiar view of earth as our own eyes see it is shown at the right in a photograph taken by the Apollo 8 astronauts from lunar orbit. *(NASA photographs 68–H–1401 [in visible light] and 72–H–761 [in ultraviolet])*

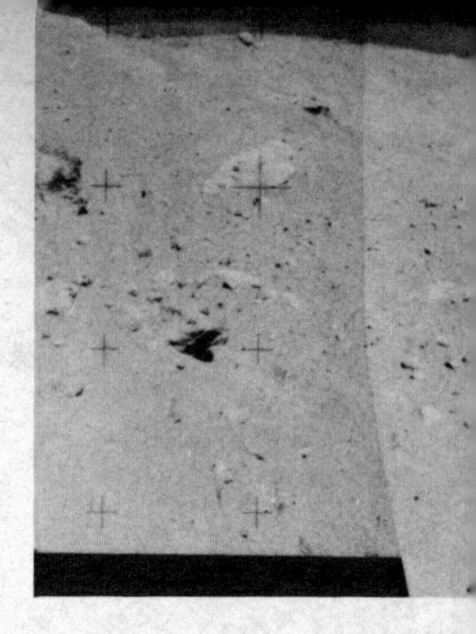

14 / A Large Meteorite Crater Samples Deep Lunar Material. North Ray Crater was a prime site for collecting lunar samples on the Apollo 16 mission. The crater, shown here in a mosaic of overlapping photographs, is about a kilometer across and probably blasted out boulders from several hundred meters below the surface when it formed. Collecting around the rim of North Ray Crater, the Apollo 16 astronauts could sample material from deep beneath them. *(NASA photograph 72–H–774)*

15 / A Sampling Problem in the Littrow Valley (left). Apollo 17 geologist–astronaut Harrison Schmitt collects small samples from a huge split boulder on the north side of the Littrow Valley. Schmitt is holding a *gnomon,* a tripod-like instrument which provides a scale and color chart for surface photographs. Behind Schmitt is the flat floor of the Littrow Valley and the steep slopes of South Massif about 8 kilometers (5 miles) away. *(NASA photograph AS17–140–21496)*

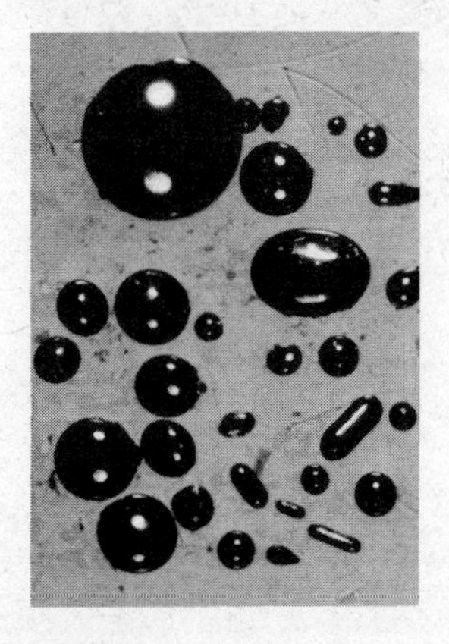

16 / Orange Soil: Young or Old? Tiny orange-brown glass droplets, the largest about half a millimeter (1/50 inch) across, make up the Orange Soil collected by the Apollo 17 astronauts from Shorty Crater in the Littrow Valley. *(Photograph courtesy of E. Roedder)*

17 / Measuring the Moon from 100 Kilometers Up. This view of the Apollo 17 Command Module *Challenger,* taken from the Lunar Module just before the two spacecraft docked, shows the scientific instruments used to photograph the moon and make chemical analyses of the moon's surface from lunar orbit. The instruments appear as box-like objects in the open space at the rear of the Command Module (top of picture). *(NASA photograph AS17–145–22254)*

18 / Building Blocks of a Lunar Basalt (top of facing page). A thin slice of a lunar basalt returned by Apollo 11 shows the interlocking crystals of different minerals that make up the rock. The rock (sample 10047) is composed almost entirely of three minerals: pyroxene (gray crystals with numerous cracks); feldspar (clear, lath-shaped crystals); and ilmenite (black opaque crystals). *(NASA photograph S–69–47907)*

19 / *The Last Bit of Liquid in a Lunar Basalt* (below). A spectacular photograph shows the glass droplets formed from the last bit of liquid remaining when over 99 percent of the rock was solid. The tiny droplets of glass (gray), less than a tenth of a millimeter across, are greatly enriched in silica (SiO_2) in comparison to the original basalt. The droplets were trapped in a crystal of the mineral pyroxferroite (a new mineral found only in lunar rocks) which forms the white background. Apollo 11 sample 10003. *(Reprinted with permission from E.L. Roedder et al.*, in Proceedings of the Apollo 11 Lunar Science Conference, *vol. 1, p. 815, © 1970 by Pergamon Press, Ltd.)*

20 / *A Lunar Microbreccia: Little Pieces of Other Rocks.* A specimen of microbreccia, about 25 millimeters (1 inch) long, contains tiny pieces of white, feldspar-rich rocks (1) probably derived from the lunar highlands. The lower part of the specimen shows several pits (2) produced by the impact of very small meteorites while the rock was on the lunar surface. Some of these pits contain small drops of glass formed by the heat of the impact. A larger splash of glass (3) coats part of the specimen at the upper left. Sample 10019, collected by the Apollo 11 mission. *(NASA photograph 70–H–233)*

21 / A Large Breccia in Molten Rock. Collected from the Descartes highlands, this 15-centimeter (6-inch) specimen contains numerous fragments of white, feldspar-rich rock in a matrix of darker formerly-molten material. The mixture of rock fragments and molten rock was probably formed by the impact of a large meteorite. Gases released from the molten liquid formed the large, contorted bubbles in the center of the rock. Sample 68815, collected by the Apollo 16 mission. *(NASA photograph 72–H–677)*

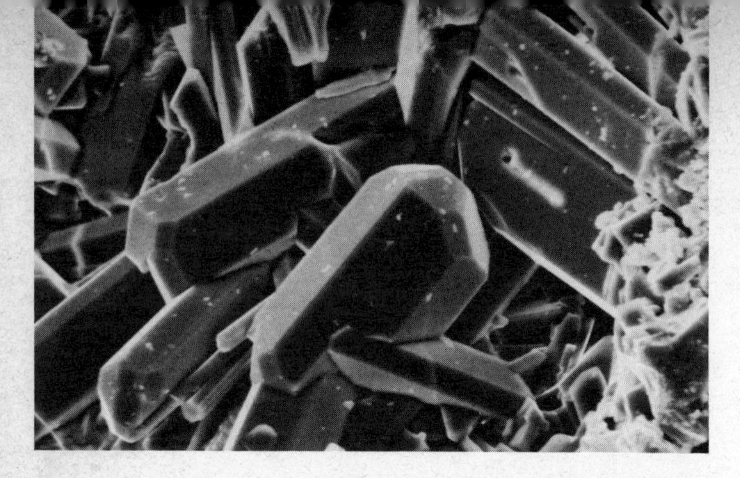

22 / *Crystals Formed from a Lunar Vapor.* Delicately perched in a small cavity in a breccia collected by the Apollo 14 mission, small crystals of the mineral *apatite* (calcium phosphate) formed directly from the vapor that filled the cavities when the rocks were formed almost four billion years ago. The perfect shapes of the crystal faces are shown in this view with an electron microscope; the longest crystal is only 0.05 millimeters long. *(NASA photograph 72–H–35)*

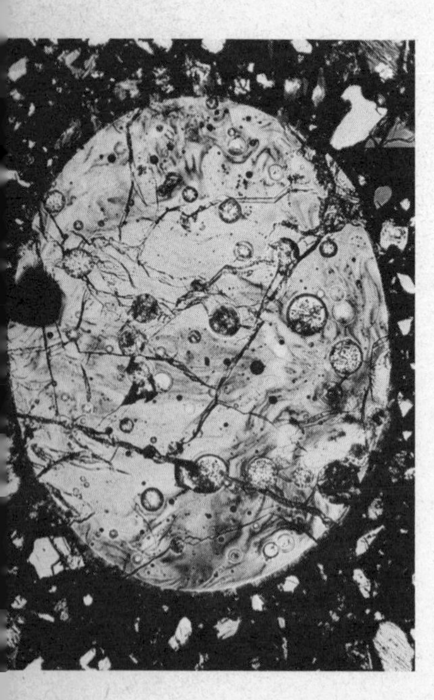

23 / *A Glass Drop from a Meteorite Impact* (left). A very small glass bead, about half a millimeter long, was formed by the sudden melting of rock struck by a small meteorite. The melted drop was ejected from the crater as part of a spray of droplets to become part of the lunar soil collected by the Apollo 11 mission (sample 10065). The bead shows irregular flow lines produced by the incomplete mixing of glasses of different chemical compositions produced by melting different minerals in the original rock. The very tiny black specks in the glass are small spherules of nickel–iron, probably parts of the meteorite that produced the melting. The droplet is surrounded by black lunar soil which contains smaller fragments of glass and broken crystals of minerals. *(Reprinted with permission from E.C.T. Chao et al., in* Proceedings of the Apollo 11 Lunar Science Conference, *vol. 1, p. 302, © 1970 by Pergamon Press, Ltd.)*

24 / A Lunar Crater on the Earth. Meteor Crater, Arizona, located about 100 kilometers (60 miles) east of Flagstaff, Arizona, is one of the best-preserved meteorite impact craters on the earth. About 1,220 meters (4,000 feet) across, it was formed by the impact of a meteorite about 30 meters (100 feet) in diameter 25,000 to 50,000 years ago. Meteor Crater was one of many geological training sites selected for the Apollo astronauts. *(Photograph from John S. Shelton,* Geology Illustrated *[San Francisco: W. H. Freeman Co., 1966], used by permission)*

25 / A Terrestrial Copy of a Lunar Rock. Glass-rich microbreccia collected from a large, ancient impact crater in Canada has many things in common with lunar impact breccias. It contains numerous crushed and shattered rock and mineral fragments and shows an intimate mixing of different kinds of light and dark glass. The long dimension of the photograph is about two millimeters. The specimen came from West Clearwater Lake, Canada, an impact crater nearly 32 kilometers (20 miles) in diameter and about 300 million years old. *(Photograph courtesy of M. R. Dence)*

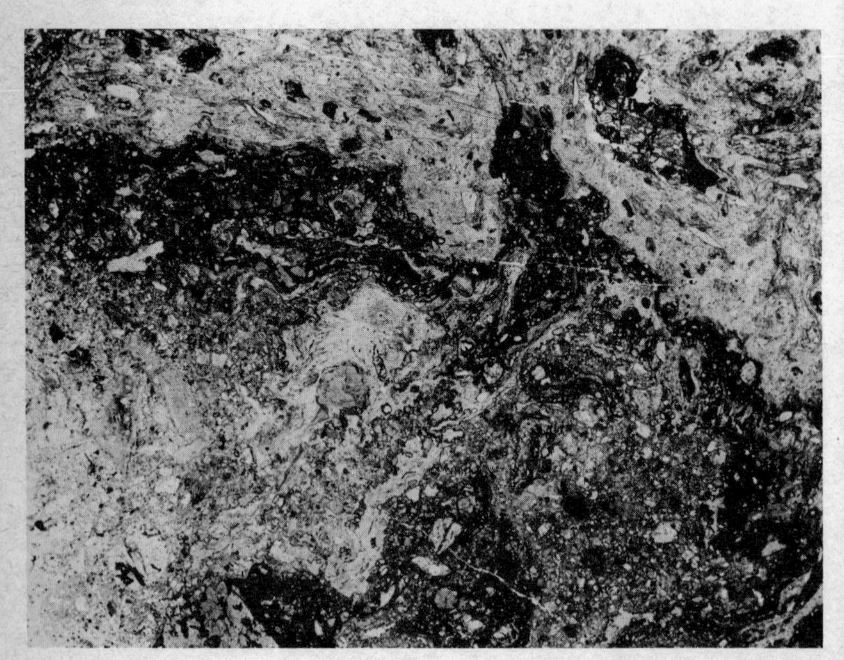

26 / The Faces of the "Microcrater Clock." The exposed face of an Apollo 16 rock sample (60015) shows an impressive assortment of microcraters ranging in diameter from about half a millimeter down to the limit of visibility. Each microcrater contains a central pit lined with dark glass (in which some light reflections can be seen) and is surrounded by a white halo of fractured rock. By measuring the age of the glass lining in each microcrater, scientists hope to develop a precise "microcrater clock" to measure the length of time that rocks like this one have been exposed on the surface of the moon. *(Reprinted with permission from H. Fechtig et al., in* Proceedings of the Fifth Lunar Science *Conference, vol. 3, p. 2464, © 1974 by Pergamon Press, Ltd.)*

27 / Microcraters on a Lunar Glass Bead. A small glass spherule from the lunar soil, only a few tenths of a millimeter in diameter, has been struck by even smaller particles which formed small impact craters on its surface. The typical microcraters have a central, glass-lined pit, surrounded by a larger ring where material has been broken off. Another microcrater is visible in profile as a gap in the edge of the spherule. *(Photograph courtesy of D.S. McKay and U.S. Clanton)*

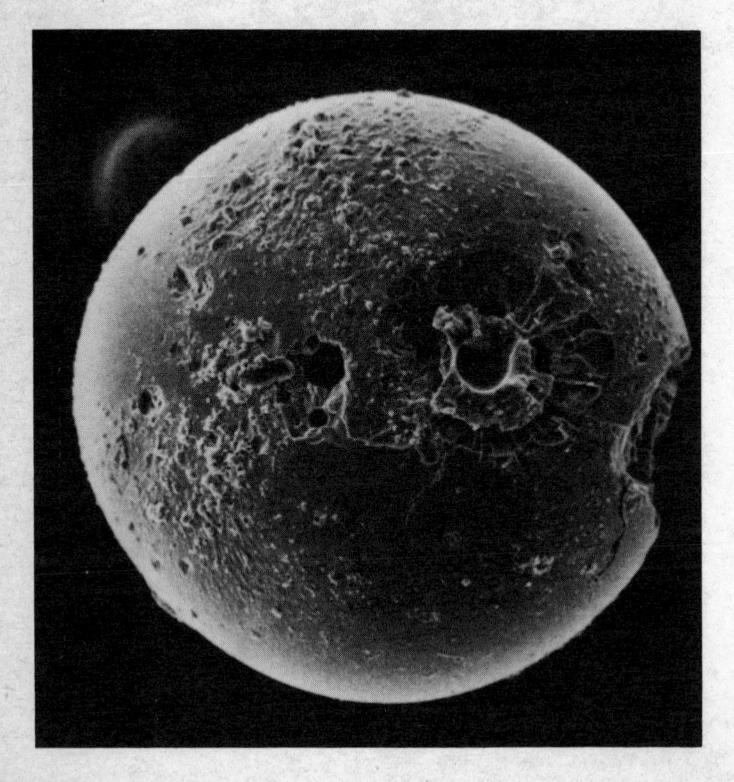

28 / "Magic Component" in the Lunar Soil. This small fragment of ropy, yellow-brown glass, less than a millimeter across, is typical of the material called KREEP, a type of lunar rock which contains unusually high amounts of potassium (K), rare-earth elements (REE), and phosphorus (P). This specimen, returned with soil samples collected by Apollo 12, is shown here in a photograph taken with an electron microscope. Chemical patterns in the KREEP material were apparently established more than 4.4 billion years ago, thus providing a partial record of very ancient lunar history. *(Reprinted with permission from D.S. McKay et al., in* Proceedings of the Second Lunar Science Conference, *vol. 1, p. 755, © Pergamon Press, Ltd., 1971)*

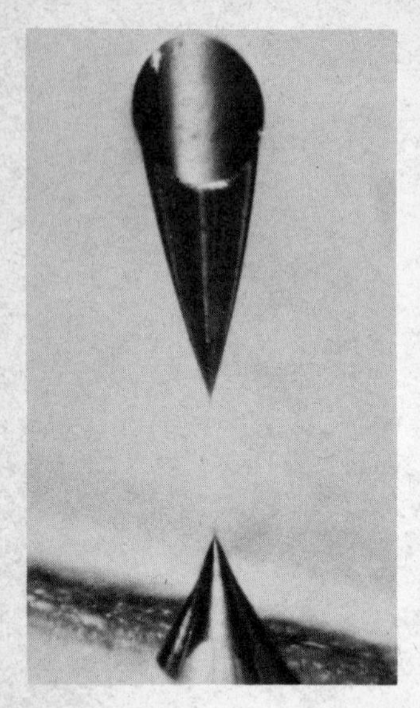

29 / *Cosmic Ray Tracks in the Apollo Spacecraft.* Cosmic rays that passed through the Apollo 14 spacecraft during its lunar trip made these permanent tracks in pieces of Lexan plastic that were part of one of the scientific experiments carried on the mission. The etching techniques used to reveal the tracks turns them into hollow tubes and cones in the plastic. The track at the upper left, about two millimeters long, consists of two cones and was probably formed by an argon atom. *(Reprinted with permission from R. L. Fleischer et al.,* Science, *vol. 181, p. 436, © the American Association for the Advancement of Science, 1973)*

30 / *Bubbles in a Lunar Lava.* This frothy-looking specimen of lava, 10 centimeters (4 inches) long, preserves a record of bubbles produced more than three billion years ago by gases trapped in molten lava. The gases, which escaped when the lava solidified, left behind a rock composed mainly of bubbles, each with a smooth, gleaming lining. Sample 15556, collected by the Apollo 15 astronauts. *(NASA photograph S–71–43328)*

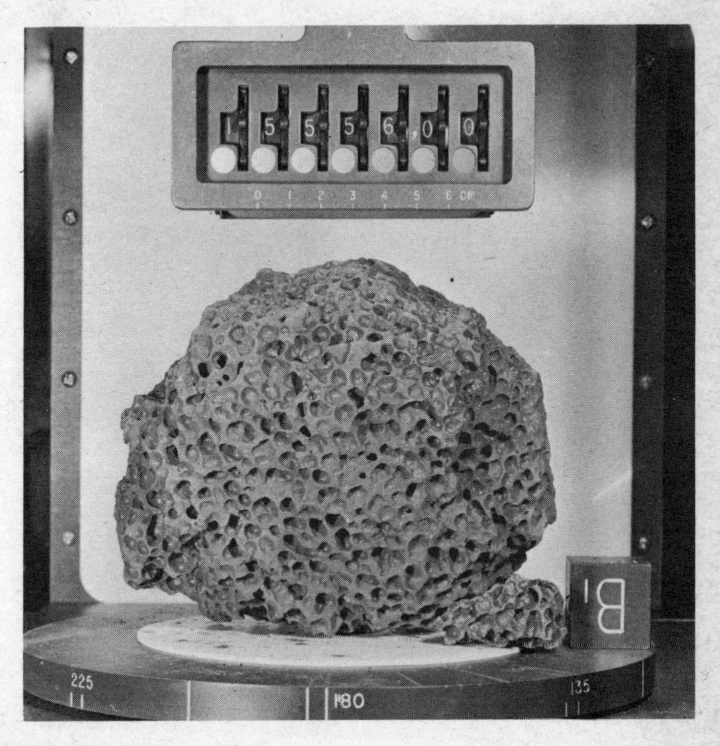

31 / A *Black, Eternally Frozen Lunar Lake.* Seen from the Apollo 15 Command Module, the crater Tsiolkovsky, 240 kilometers (150 miles) in diameter, on the back side of the moon, displays its high, white central peak and dark, flat floor of mare material. Tsiolkovsky is one of the few scattered areas of dark mare material on the back side of the moon, whereas nearly half of the front side is covered with mare material. *(NASA photograph AS15–91–12383)*

32 / *A Farewell Look at a Brand-New Moon.* Most of the formerly unknown back side of the moon is clearly shown in this view taken by the Apollo 16 astronauts soon after starting their return to earth. *(NASA photograph 72–H–848)*

33 / *Mercury at Last Quarter.* Mercury, the closest planet to the sun, presents a heavily cratered, moon-like appearance to the cameras of the unmanned spacecraft Mariner 10. This panorama of Mercury is made from 18 pictures taken on March 29, 1974, when Mariner 10 was about 200,000 kilometers (125,000 miles) from the planet. Most of the sunlit part of Mercury is in the planet's southern hemisphere. The diameter of Mercury is about 5000 kilometers (3100 miles), and the largest craters shown are about 200 kilometers (125 miles) in diameter. *(NASA photograph 74–H–39)*

34 / A Gigantic "Mount Olympus" on Mars. Olympus Mons, which may well be the largest volcanic mountain in the solar system, was photographed by Mariner 9 in January, 1972, as a Mars-wide dust storm subsided. The giant peak, gradually revealed in the clearing atmosphere, is 500 kilometers (310 miles) in diameter and rises about 25 kilometers (15 miles, or three times the height of Mt. Everest) above the surrounding plain. *(NASA photograph 73–H–104)*

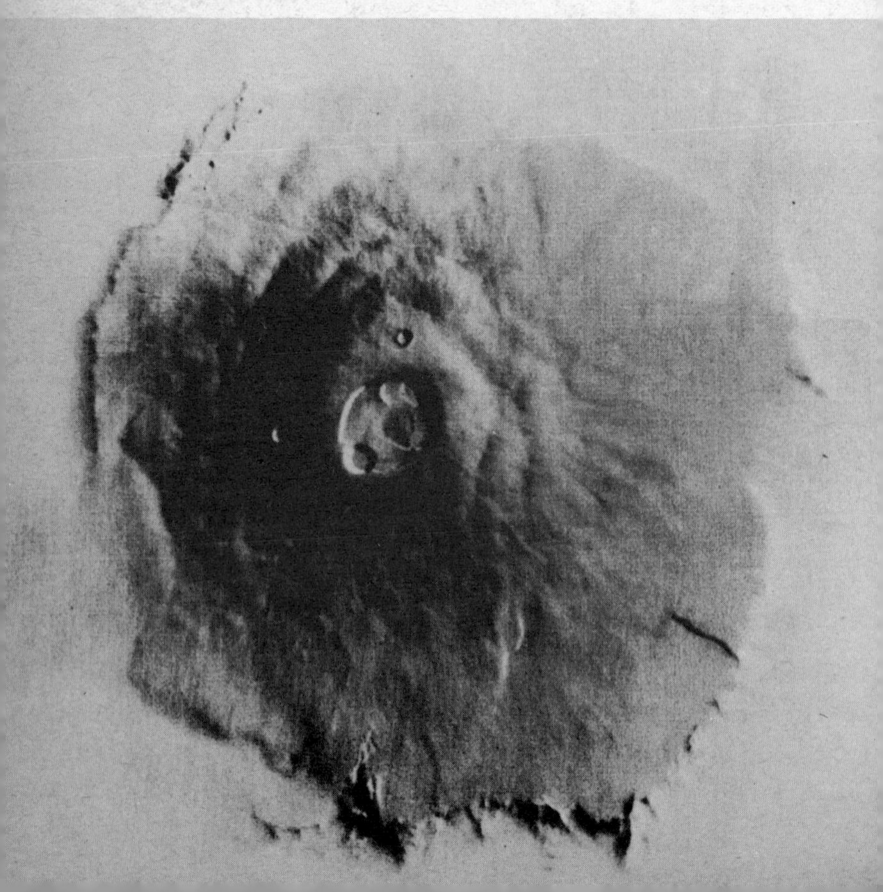

35 / "*. . . To See Ourselves as Others See Us.*" Robert Burns's poetic wish is fulfilled by the returning crew of Apollo 17 in this spectacular photograph of the earth. Madagascar and the east coast of Africa occupy the center of the picture. *(NASA photograph 72–H–1578)*

36 / "We Came in Peace for All Mankind." With the last explorations of the Apollo Program nearly completed, Apollo 17 astronaut Eugene Cernan stands on the surface of the moon between the Lunar Rover and the American flag. *(NASA photograph 73–H–199)*

CHAPTER

ANATOMY OF A MOON ROCK

As the last Apollo mission returned safely to earth, the scientific focus of the program shifted to dozens of laboratories where scientists were working hard to determine what the lunar rocks were made of, how they had formed, and how old they were. Eventually the picture of the moon provided by astronauts and instruments would be combined with the record of the past that had been frozen in the rocks to provide a continuous picture of what the moon had been like in the past and how it had become what it is today.

When the first lunar samples were returned to earth, both the rocks and the Apollo 11 astronauts themselves went immediately into quarantine for three weeks. The rocks were carefully examined for microscopic life, and test animals and plants were exposed to the rocks and injected with lunar dust. Doctors examined the astronauts to see if they had picked up alien diseases from any lunar germs. The astronauts, to everyone's relief, remained in fine health. The mice, quail, fish, oysters, insects, and plants that were exposed to the lunar material showed no ill effects. No microorganisms grew in the cultures made from lunar rocks. Analyses of lunar rocks found no organic compounds and only tiny amounts of inorganic carbon. Just like the observable surface of the moon, the lunar rocks were devoid of life. The quarantine period ended without incident, and the Apollo 11 samples were distributed to waiting scientists. Eventually, the sterility of lunar rocks became so accepted that the quarantine was completely dropped after the Apollo 14 mission.

The Apollo samples became the center of a program of scientific research that had been carefully planned for several years before the first lunar landing. About 1,000 scientists in the United States and many foreign countries participated in the analysis of the lunar samples. Their research drew on nearly a century of scientific experience. The methods and instruments that had proved so successful in explaining the nature and origin of earth rocks were now applied to samples of the moon.

The Apollo samples were analyzed with the newest and most advanced equipment available. A battery of complex instruments measured nearly every physical and chemical property that could be determined: chemical composition, density, radioactivity, thermal conductivity, electrical conductivity, magnetism, and many others. Despite all this modern technology, one of the favorite instruments for the geological study of the lunar samples had hardly changed at all since it was first developed for use on terrestrial rocks more than a century ago: the petrographic microscope.

A petrographic microscope is similar to a conventional microscope except that it uses polarized light to determine the microscopic textures of rocks and to identify the minerals in them.* There is even a separate branch of geology

* Polarized light is created by passing a beam of ordinary light through a transparent crystal. The original beam consists of many light waves, each of which moves up and down in a single plane.

called petrography that deals with the study of rocks with such microscopes. These instruments have long been standard equipment for studying terrestrial rocks, and they were applied to the new moon rocks without any modifications.

To study a rock microscopically, a small chip of the sample is first cemented to a glass slide. Then the chip is ground down and polished until it is thinner than a sheet of paper (0.03 millimeters, or about one-third the thickness of a human hair). The sample is then so thin that light can pass through it. Looking through the microscope at this sample (called a "thin section") an observer can identify fine textures and tiny crystals that could not be seen by the unaided eye. Under the microscope, the crystals in the rock act as small polarizing prisms, and they produce a beautiful play of colors that is used to identify the different minerals in the rock.

The great value of this microscope, even in an age of complicated and sensitive analytical instruments, is that it provides a direct look at how a rock is put together. Such features as the sizes and shapes of crystals and the relationship of one crystal to another tell much about the origin and history of the rock. With a quick look through a microscope, a scientist can determine how the rock formed, whether it was once a liquid, whether it crystallized quickly or slowly, whether one mineral formed before another, or whether the rock was broken up or melted again after it had formed.

In ordinary (unpolarized) light, these planes are oriented in all directions, but as the beam passes through the crystal, its atomic structure absorbs all the light waves that are not oriented in a single direction. The light that emerges from the crystal contains only waves that vibrate in the same plane; such light is called polarized.

Because polarized light is produced by the internal structure of crystals, it is a powerful tool for investigating and identifying transparent minerals. In a laboratory petrographic microscope, polarized light is produced by passing light through a specially prepared crystal of the mineral calcite (Iceland Spar). The same effects can be produced with commercial Polaroid, a material composed of very tiny crystals all arranged in the same direction.

Chemical analysis was an equally important part of the study of lunar rocks. Methods of chemical analysis had been developed and used on terrestrial rocks for many years. Analysis begins by dissolving a sample in acid, followed by chemical tests on the solution to determine the amount of each element present. The problem with using such a technique on the lunar rocks is that it uses up too much material, about 5 grams for each analysis, or an amount equal to about a dozen aspirin tablets. This amount is insignificant for terrestrial samples when one can collect kilograms, or even tons, of material. But lunar material is scarce, and there were many scientists waiting to receive samples.

Fortunately, a new instrument for chemical analysis had been developed just a few years before the Apollo 11 landing: the electron microprobe. This machine, which was just beginning to be used to analyze terrestrial rocks, had several advantages over older methods that made it ideal for analyzing lunar samples. First, the instrument was nondestructive, so that analyses could be made without using up scarce material. Second, the analyses could be made on the same thin sections that had been prepared for microscopic study; thus a great deal of information might be obtained from as little as a tenth of a gram of lunar soil or from a piece of rock the size of a pinhead.

The electron microprobe is a complicated instrument that fills a small room, but it operates on simple principles (*Figure E*). It contains a source of electrons similar to the "electron gun" in a TV picture tube; this device bombards the sample with a stream of high-energy electrons. The electrons excite the atoms in the sample and cause them to give off X-rays. The rest of the electron microprobe consists of components which detect and record the X-rays. Each element in the sample has a distinctive X-ray, and from the intensities of the different X-rays, the amount of each element present can be calculated. Usually, the electron microprobe is attached to a small computer which can calculate

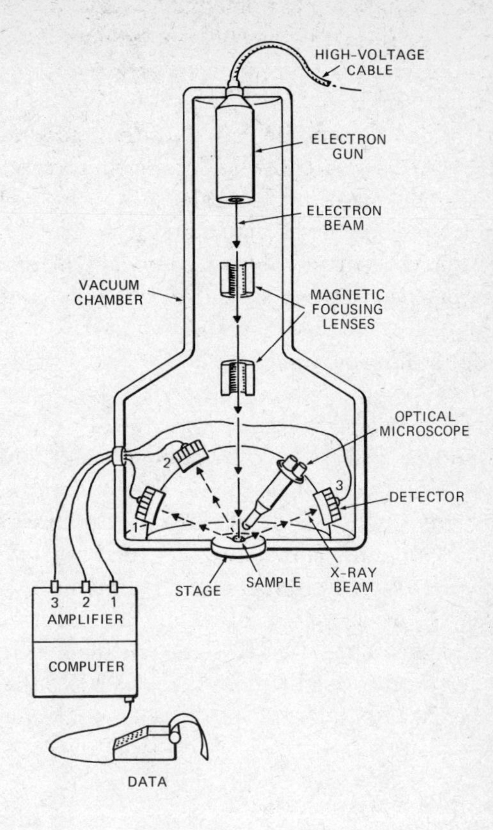

Figure E / Chemical Analyses the Easy Way. The electron microprobe, shown here in a simplified diagram, is essential for obtaining accurate and nondestructive analyses of small lunar samples. A beam of electrons (solid line) is generated at the top of the instrument and then focused by magnetic lenses onto the lunar sample at the bottom. The electron bombardment causes the atoms in the sample to emit X-rays (dashes) that are picked up by the detectors. From the intensity of the different X-rays, the chemical composition of the sample can be computed.

the chemical composition immediately. A complete chemical analysis for 10 to 12 elements can be done in 2 minutes. Using the old methods, the same analysis would have taken several days.

The electron microprobe has a further advantage. The electron beam is so narrow and so well focused that an analysis can be made on a single spot only a few thousandths of a millimeter in diameter. As a result, the scientists working with lunar samples could obtain hundreds of chemical analyses from a single crystal only a few millimeters long. They could measure variations in chemical composition within a single crystal and thus obtain important information about how the crystal had grown. They could detect and analyze crystals too small to be seen with the eye, and they were able to discover and analyze new minerals that had never been found in terrestrial rocks.

Of the tens of thousands of chemical analyses of lunar rocks, soils, minerals, and glasses, virtually all were made with the electron microprobe. Without this instrument, our knowledge of the chemistry of the moon would have been uncertain, incomplete, and lacking in fine details. The Apollo spacecraft carried men to the moon, but the electron microprobe helped make it possible to understand what they brought back.

THE ROCKS

Only a few days after the Apollo 11 samples were unpacked, the first preliminary analyses established that the moon rocks were similar to terrestrial rocks in their general chemistry. Scientists found that the earth and moon were made out of the same chemical ingredients, although the proportions were slightly but significantly different. Like the earth, the moon turned out to be very different chemically from the sun, which contains 99 percent of the matter in the solar system. The sun contains 89 percent hydrogen, whereas the earth has very little (only enough to form the water in its oceans); the moon apparently has none.

Although the basic chemistry of the earth and moon is similar, there are important differences between earth rocks and moon rocks in their appearance and origin. On the earth, geologists have long recognized three basic kinds of rocks. Igneous rocks, such as granite and basalt, form by the cooling and solidification of molten silicate liquids that erupt from volcanoes or solidify as huge masses below the earth's surface. Sedimentary rocks, such as sandstone, shale, and limestone, are formed out of layers of material deposited in the oceans. Metamorphic rocks, such as gneiss, schist, and marble, are formed when igneous or sedimentary rocks are buried deep in the earth and metamorphosed into new rocks by high temperatures and pressures.

In contrast to the three types of rock found on Earth, all the Apollo samples were identified as igneous rocks that had solidified from a once-molten material. No sedimentary rocks like shale or limestone were found. However, the Apollo samples were not all the same in appearance. Even the unaided eye could see that different moon rocks had different minerals and textures. Different kinds of igneous rocks were found in different parts of the moon, and scientists realized that the geology of the moon, while lacking some features of the earth's geology, would still be a complex and challenging problem to unravel.

LAVAS FROM THE MARIA

The first samples returned from the lunar maria by Apollo 11 and later missions turned out to be fairly simple rocks, and they were quickly identified as a type of lava called basalt. Basalt is a common rock on earth. It forms the lavas that erupt spectacularly from volcanoes in Hawaii and Iceland and which pour out more quietly from deep fissures in the ocean floor. About 15 million years ago, large floods of basalt built up the wide, level expanse of the Columbia River Plateau in the northwestern United States.

The similarity between the lunar rocks and terrestrial basalt lavas was a fortunate, if somewhat unexpected, discovery; now geologists could draw on their experience with terrestrial basalts to explain the origin of the lunar maria. They argued by analogy that the maria had been built up gradually by successive eruptions of molten lava that poured out and then spread for tens, or even hundreds, of kilometers across the lunar surface. Like lavas on earth, most individual lunar lava flows were only a few meters thick, but one eruption had followed rapidly on another, and eventually hundreds of flows had built up a pile of layers many kilometers thick that filled the mare basins and spread across the lunar "seas." Some of these individual layers were seen by the Apollo 15 astronauts in Hadley Rille, where the Rille cut a cross section through a stack of flows. Elsewhere on the level maria the layers themselves were hidden under the surface deposit of lunar rubble.

Geologists have established that the lavas that erupt today from earth's volcanoes have formed from the melting of part of the earth's interior by its internal heat. The melting takes place 50 to 200 kilometers deep within the earth, and the molten lava makes its way to the surface through cracks and deep fissures. The discovery of lavas on the lunar maria indicated that, about 3½ billion years ago, the moon had behaved much as the earth does today. It was reasonable to conclude that the moon must have been hot inside—hot enough to have melted, thus forcing molten lava to rise from deep within the interior and erupt over its surface.

The identification of the maria rocks as basalt flows was based chiefly on the microscopic examination of their textures and the identification of the minerals that composed them. Under the microscope, the lunar samples showed the same textures of interlocking crystals typical of terrestrial rocks formed from a cooling melt (photo 18). Further study showed that the lunar basalts were composed almost entirely of the same minerals that are found in terrestrial basalts.*

Pyroxene, a Ca–Mg–Fe silicate, is the most common mineral in lunar lavas, and it makes up about half of most specimens. It forms yellow-brown crystals that are several millimeters to a few centimeters in size and often large enough to be seen with the naked eye. Olivine, an Mg–Fe silicate, is a less common companion of pyroxene; it may be absent from many lunar lavas. When it occurs, it forms pale green crystals a few millimeters in size. Plagioclase (or feldspar), a Ca–Al silicate, forms elongated white crystals, and is colorless and transparent in thin sections. Ilmenite, an Fe–Ti oxide, is also common; it occurs as black-bladed crystals that are opaque even in thin sections. Spinel is an Mg–Fe–Cr–Al oxide that occurs in small amounts in lunar lavas, usually as small black crystals.

Despite the many similiarities between lunar and terrestrial basalts, there are important differences between them, and a geologist can easily distinguish a piece of terrestrial basalt from a lunar one. For one thing, the lunar rocks contain no water, while terrestrial basalts always have 1 percent or so. To the layman this may seem like a small point, but it is a critical fact in understanding the different histories of the earth and moon. To the geologist looking through a microscope, it means that moon rocks look much

* Minerals are naturally occurring chemical compounds; rocks are composed of one or more minerals. The minerals most common to igneous rocks from both the earth and moon are called silicates, and they are composed chiefly of the chemical elements silicon (Si) and oxygen (O). Other elements that commonly occur in silicate minerals are calcium (Ca), aluminum (Al), iron (Fe), magnesium (Mg), sodium (Na), and potassium (K). Most igneous rocks also contain oxide minerals, which are composed of oxygen and metals such as iron, titanium (Ti) and chromium (Cr).

fresher than earth rocks, which have been more or less altered by water ever since they were formed. The moon rocks contain no water-bearing (hydrous) minerals, no clays, and no rusty iron oxides. Geologists were startled to discover that a lunar basalt 3½ billion years old actually looked fresher than warm lava collected from a terrestrial volcano only a few days after an eruption.

I think it was this freshness that impressed me most the first time I looked through the microscope at a lunar basalt. An excited colleague, who had just received his first thin section of a lunar rock, dashed into my office and shoved the tiny glass slide literally under my nose and onto the stage of the microscope I was using at the time.

The mosaic of crystals that the microscope revealed was familiar. Pyroxene crystals, pale yellow-brown, were intergrown with transparent laths of plagioclase feldspar and the black plates of ilmenite. A touch of a lever brought in the polarizing prism, and the view took on the colors and patterns of an abstract stained-glass window. The pyroxenes shone out in bright reds and blues and yellows, while the plagioclase crystals took on muted grays and whites and showed a pattern of parallel stripes that revealed the finer details of their crystal structure.

Even without the polarized light, the colors within single pyroxene crystals were variable. They were pale yellow-brown in the centers, and the color changed to a deeper cinnamon tint at the edges. The electron microprobe later showed that the darker rims were rich in iron and titanium.

The textures and the minerals were familiar. What was strange was the lack of alteration—the clearness and freshness of every crystal. In even a fresh terrestrial lava, the pyroxene grains would be cut by tiny veinlets of green minerals such as serpentine or chlorite that had been formed by the reaction of water with the pyroxene. The plagioclase crystals would be spotted with flakes of clay minerals, and the ilmenite would have developed a whitish coating of iron and titanium oxides. No such alterations were found in the lunar rocks. The crystals were unspotted, and even the patches of glass that sometimes occurred in the spaces between grains were clear and unchanged. No rock like this could ever have formed on earth.

The chemical compositions of the lunar lavas showed some important differences from terrestrial basalts. The lunar basalts contained more titanium, although the amount of that element varied widely, from about 3 to 5 percent TiO_2 in the Apollo 12 samples to over 10 percent in the Apollo 11 samples. (Most terrestrial basalts contain only about 1 to 3 percent TiO_2.) The lunar basalts also had about twice as much iron as terrestrial basalts, and virtually all the lunar iron occurs as the less oxidized (or ferrous) oxide, FeO, instead of the more oxidized (ferric) form, Fe_2O_3, that accompanies FeO in terrestrial lavas.

The less oxidized character of the iron in lunar rocks suggested that they had formed under conditions where there was practically no free oxygen available. This idea was confirmed by the discovery of small amounts of iron metal in the lunar lavas.

The lunar minerals provided some exciting and unexpected details about the low amount of free oxygen present while the lavas were cooling. When a terrestrial lava cools, the pressure of free oxygen present in the rock is less than a hundred-millionth of the pressure of oxygen in the earth's atmosphere. (Most lavas, even at the surface, are so solid that atmospheric oxygen does not penetrate them while they are cooling.) However, even this low amount of oxygen is usually enough to oxidize the iron in the lava to form oxide and silicate minerals, leaving no iron in its free metal state.

The discovery of small amounts of metallic iron as a mineral in the lunar lavas indicated that there was even less free oxygen available when the lunar rocks had formed. The iron and other minerals in the lunar rocks told geologists that the oxygen pressure in lunar basalts was only one hundred-thousandth that in terrestrial ones, and the moon appeared to have less free oxygen available than even the inside of the earth.

Other chemical elements were less abundant in the lunar rocks than in terrestrial ones, and these relative scarcities also gave important information about the moon. In addition to the absence of water, the lunar rocks had only about one-tenth of the sodium (Na) and potassium (K) found in terrestrial lavas. Water and the alkali elements, as sodium and potassium are called, have one important thing in common: they are all volatile substances, that is, they are easily boiled out of a rock by heating it to high temperatures. The lack of these materials in lunar rocks suggested that the solid matter that formed the moon had been extremely hot, possibly above 2,000° C., at some time in the moon's history. During this early heating the volatile substances (like water and the alkali elements) were boiled off the moon and lost into space. Despite the many chemical similarities between lunar and terrestrial rocks, here was evidence that the history of the moon had been very different from the history of the earth.

These results were especially discouraging to those who had hoped to find lunar life. None had yet been found, either on the lunar surface or in the rocks themselves. And now the rocks were telling us that the moon had no water and probably no free oxygen, two substances essential to life. The moon has no life now, and the chemical data from the rocks make it unlikely that it ever had any.

The different chemistry of the lunar rocks also affected their behavior when they were first erupted. Because the lunar lavas contained more iron and titanium, they were more fluid than earth lavas. The molten rock on the surface of the moon was about as fluid as lubricating oil, and thus a single lava flow could spread rapidly across the moon for hundreds of kilometers before solidifying.

Both lunar and terrestrial basalt are mixtures of various

chemical compounds, all of which have different melting points. Thus, molten basalt liquid does not freeze all at once at a single temperature the way a pure substance like water does. Instead, a cooling basalt liquid solidifies gradually over a range of temperatures, and different minerals crystallize from the liquid at different times as the temperature drops.

By comparing the chemical compositions of individual lunar minerals with the results from many years of laboratory experiments, scientists could estimate the temperatures and pressures at which the lavas had solidified. When the lunar lavas were first erupted and began to cool and crystallize, their temperature was about 1,200° C. (2,200° F.)—about the same temperature at which basalt lavas are erupted on the earth. As the lunar lavas cooled, more and more crystals formed from the liquid, but the rocks did not become completely solid until the temperature had dropped to about 950° C. (1,750° F.).

To the physical and chemical information, the geologist could add the perspective of microscopic examination. From the textures in the rocks, he could trace the steps of their transformation from a hot, molten liquid into solid rock. The sizes and shapes of the crystals indicated that the rocks had formed by relatively rapid cooling. Geologists estimated that the lava flows had solidified completely in less than a few years. This meant that the lava flows had to be less than a few tens of meters thick, because thicker layers would have taken thousands or millions of years to solidify completely.

The preserved textures demonstrated how rapidly the crystals had grown from the original liquid. Through the microscope, scientists observed crystals that had grown so fast that the surrounding liquid did not have time to move out of the way. Many droplets of liquid were actually surrounded and enclosed by the growing crystals (photo 19). These isolated bits of liquid then cooled and crystallized without any further communication with the rest of the lava. Some of these trapped droplets were only a fraction of a millimeter in size. Others were large enough to form microscopic crystals of their own, and these larger bodies gave scientists a chance to study different patterns of crystallization in the same rock and sometimes in the same crystal.

The first crystals to form in a cooling lunar basalt are usually olivine or pyroxene. These minerals contain a higher percentage of calcium and magnesium than the original basalt lava. The formation of these minerals thus removes calcium and magnesium from the liquid. As these minerals continue to form, two things happen: the amount of remaining liquid is progressively reduced; and the liquid becomes increasingly depleted in calcium and magnesium and correspondingly enriched in all the other elements.

This process, called "fractional crystallization," causes the composition of the remaining liquid to change continuously as crystals form. It is very similar to the common process of freezing salt water; some of the water freezes out as pure ice crystals, while the salt becomes more strongly concentrated in the remaining liquid brine.

Through the process of fractional crystallization in the lunar lavas, many chemical elements that were uncommon or even rare in the original basalt liquid are strongly concentrated in the small amount of liquid that remains when the rock is more than 95 percent solid. At this point, the compound silica, SiO_2, becomes so enriched in the remaining liquid that minerals composed of pure SiO_2, called tridymite and cristobalite, are formed. Even rarer minerals, containing large amounts of such elements as potassium, zirconium, sulfur, and radioactive uranium and thorium are also formed.

A number of new minerals, never observed on earth, were also found in the lunar rocks. These minerals formed because of the unusual chemical conditions that existed during the cooling of the lunar lavas, especially the absence of water and the unusually low oxygen pressure. Many of the minerals were found as tiny crystals in cracks and veinlets, where they had crystallized from the last remaining liquid in the rock. Most of these crystals were so tiny and so chemically unusual that they could be detected and identified only with the electron microprobe.

One new mineral, which occurs in clear yellow crystals, is similar to pyroxene, but it is enriched in iron because of the unusually low oxygen pressure in the lunar rocks. Its name, pyroxferroite, derives from its similarity to pyroxene and from the chemical symbol for iron (Fe).

A more common new mineral is an opaque oxide of iron, titanium, and magnesium somewhat like ilmenite. Its name, armalcolite, combines the initial letters from the names of the Apollo 11 astronauts (*Arm*strong, *Al*drin, and *Col*lins). It was found in the first samples brought back by Apollo 11, and it was a common mineral in the lavas returned by Apollo 17 from the Littrow Valley a thousand kilometers away.

A third new mineral occurs as tiny orange-red crystals in both the Apollo 11 and Apollo 12 basalts. Called tranquillityite, after the lunar mare where it was first collected, the mineral is composed mostly of iron, titanium, and silicon, but it is unusually enriched in the rare elements zirconium, yttrium, and uranium. The process of fractional crystallization has so concentrated these last elements into the tranquillityite that the crystals contain more than a hundred times the amounts present in the original basalt.

No matter where the Apollo missions sampled the lunar maria, the rocks all testified to the same simple process of formation—solidification from molten lava. The minor differences which existed between one mare and another did not contradict this general conclusion. The Apollo 12 basalts, for instance, had less titanium and were nearly 400 million years younger than the Apollo 11 basalts, but they had formed in the same way. In fact, the mare lavas were so similar from one landing site to another that scientists could be confident that all the dark regions on the moon were made up of basalt lavas.

The lavas from the maria provided a picture of the moon that differed greatly from the pre-Apollo idea that the moon was a cold and primordial object. The existence of basalt lavas meant that the moon had been hot enough inside to melt. The age differences in the lavas suggested that the moon had remained hot and partly molten for almost half a billion years. The chemical differences in the lava indicated that equally significant chemical variations existed in the lunar interior; each batch of lava seemed, so to speak, to have come out of a different pot of molten rock.

We were fortunate to have landed first on the lunar maria, where the ground was flat and the rocks fairly simple. The study of the rugged lunar highlands turned out to be a more difficult matter. The rocks were more complicated, and in all of the highland samples, the record of crystallization from molten lavas was blurred and confused by catastrophic episodes of shattering, mixing, and remelting.

ROCKS FROM THE HIGHLANDS

The lunar lavas from the maria contained a detailed record of the history of the moon for the period between about 3.7 and 3.3 billion years ago. These lavas, however, formed only a thin skin that covered about one-fifth of the moon's surface. Furthermore, scientists believe that the time spanned by the lavas—about 400 million years—is less than one-tenth the age of the moon. The maria rocks provided no information about lunar events more than 3.7 billion years old.

To find out about the moon's early history, and to learn what most of it was made of, later Apollo missions landed in the lunar highlands. The highlands are the oldest part of the moon that we can distinguish from earth. In these light-colored, heavily cratered regions, which cover four-fifths of the moon's surface, is contained the record of what happened to the moon during its earliest years, long before even the oldest mare lavas appeared.

The Apollo 11 landing in Mare Tranquillitatis had also accidentally provided scientists with samples from the highlands. Although the nearest highland regions are over 50 kilometers from the Apollo 11 landing site, some tiny white fragments of plagioclase-rich rock were picked out of the samples of lunar soil collected by the astronauts. Scientists suggested that these fragments were highland rocks that had been blasted into the maria by meteorite impacts. This identification was confirmed when larger specimens of highland rocks were returned from the Apennine Mountains by Apollo 15, from the Descartes Plateau by Apollo 16, and from the mountains around the Littrow Valley by Apollo 17.

Like the mare lavas, the highland rocks are all igneous, and they, too, were formed by the cooling and solidification of molten rock. But their chemistry and their mineral composition are quite different from the mare samples. The highland rocks contain nearly twice as much calcium and aluminum as the mare lavas, and they have correspondingly less titanium, magnesium, and iron. The analysis of the rocks thus agreed with the analyses that were made from lunar orbit (see pp. 112–13), confirming that all the highlands were enriched in calcium and aluminum.

The chemical differences between highland and mare rocks are reflected in their mineral composition. Highland and mare rocks contain the same minerals, but the concentrations are different. The highland rocks, containing more calcium and aluminum, have much more plagioclase (the

Ca–Al silicate mineral) than do the mare lavas. Plagioclase usually makes up at least half of the highland rocks, occurring with varying amounts of pyroxene, olivine, and spinel.

Because highland rocks contain the same minerals as terrestrial rocks, the names of terrestrial rocks could be used to describe them. Most highland rocks are called *gabbro*, a term used for a rock that contains 50 to 75 plagioclase and 50 to 25 percent of other minerals. A more unusual highland rock, called *anorthosite*, contains over 90 percent plagioclase, and some specimens may be pure plagioclase with no other minerals. The Genesis Rock collected by the Apollo 15 astronauts and many of the rocks returned by the Apollo 16 mission were anorthosite. Any rocks with amounts of plagioclase between 75 and 90 percent are usually given hybrid names like anorthositic gabbro or gabbroic anorthosite, depending on the amount of plagioclase and the preference of the geologist doing the naming.

Despite the major chemical differences between highland and mare rocks, there are also some important similarities. The highland rocks also contain no water and practically no sodium or potassium. These results told scientists that the loss of volatile materials was a general feature of all lunar rocks and not the result of a separate process that had affected only the mare lavas. This information strengthened the view that all volatiles had been lost from the moon very early in its history. Perhaps they were boiled away from the small particles of solid matter that made up the original dust cloud long before the moon itself was formed. The highland rocks also contained small grains of metallic iron, indicating that conditions of unusually low oxygen pressure existed both in highland and mare regions.

Although the highland rocks and mare lavas both formed in the same general way—by the cooling and crystallization of molten liquids—they appear significantly different under the microscope. The mare lavas are made up of small crystals, and their textures indicate relatively rapid cooling. The crystals seen in the highland rocks are larger and more evenly shaped, and the textures indicate slower cooling and a much longer cooling time.

The mare rocks most closely resemble terrestrial lavas which erupt from volcanoes at the earth's surface and solidify rapidly within a few months or years. The highland rocks, on the other hand, are similar to other terrestrial rocks that form by the much slower cooling of larger masses of molten material deep within the earth. If the highland rocks formed deep within the moon where heat loss was slow, they may have taken millions of years to become completely solid. Once formed, they may have been forced to the surface by great upheavals in the lunar crust or blasted out by large meteorite impacts.

The ages so far measured on highland samples all fall in the range of 4.0 to 4.2 billion years, thus confirming the view that the highlands are older than the maria. However, the highland rocks are still younger than some meteorites which have fallen to earth and been dated at 4.6 billion years. Scientists think that the moon formed at this earlier time with the meteorites and the rest of the solar system. If this is true, then the first half-billion years of the moon's history has not yet been found, not even in the highland rocks.

Opinions differ on what this gap in the lunar record might mean. One possibility is that with so few Apollo landings we have not been able to find the oldest available lunar rocks. Another possible theory is that the moon remained molten all through the period between 4.6 and 4.2 billion years ago, so that no solid rocks formed during this period. A third explanation is that the moon became solid early in its history, but the highlands formed only about 4.2 billion years ago by a major episode of remelting that affected the whole moon. A final possibility is that the highlands formed very early in the moon's history but were so shattered and remelted about 4 billion years ago that no older rocks survived.

We do not know yet which explanation is correct. The period of half a billion years, which separates the formation of the moon and the solar system from the apparent ages of the oldest highland rocks, still remains almost a total mystery to us.

BRECCIAS: ROCKS FROM ALL OVER THE MOON

The initial event in the history of both mare and highland rocks is their formation by the solidification of molten lava. After they solidified, however, the lunar rocks were acted on by various forces that broke them up and mixed the pieces together to form a new variety of rock, called *breccia*.

Breccia is a name long used to describe terrestrial rocks that are made up of fragments of older rocks. On earth, breccias often indicate the shattering of existing solid rocks by sudden and violent events, such as volcanic eruptions or sudden earth movements associated with earthquakes.

Although they are most common in the highlands, breccias have been collected at every landing site on the moon. They are complicated rocks that are made up of shattered, crushed, and sometimes melted pieces of other lunar rocks. The individual pieces in a breccia may range from tiny particles less than a millimeter in size to huge blocks several meters long. The fragments may be igneous rocks of many different types and ages, blobs and droplets of glass, or pieces of other breccias.

The discovery of breccias on the moon indicates the existence of violent mechanisms that acted on both highland

and mare rocks after they had formed. Two possible processes for forming breccias on the moon are meteorite impacts and volcanic eruptions, but most scientists now feel that nearly all the breccias that have been collected have been formed by the impacts of both small meteorites and large asteroids.

The breccias returned from the maria, especially by the Apollo 11 and 12 missions, were found as rare, scattered clods that were fragile and crumbled easily when handled. Under the microscope, they are complicated rocks that contain pieces of the lava bedrock, broken-up grains of minerals, and small droplets and irregular fragments of glass of all colors—clear, green, brown, and orange (photo 20). The fragments are weakly cemented together by fine dust and glass. These breccias seem to have formed when small meteorites struck the lunar surface and compacted the loose soil into small clods that were scattered in and around the resulting crater.

The breccias in the highlands occur on a much greater scale. Here they form huge blankets that may be several kilometers thick and spread over thousands of square kilometers of the moon's surface. The astronauts who landed in the highlands collected almost nothing but breccias, and it is likely that even the few specimens of solid igneous rocks picked up were fragments from the breccias rather than nearby bedrock.

The Fra Mauro Formation, sampled by Apollo 14, and the Cayley Formation, on which Apollo 16 landed, are now interpreted to be two widespread layers of breccia that were deposited by the huge impacts that formed the basins of Mare Imbrium and Mare Orientale.

At the Apollo 14 site several different kinds of breccia were collected. One type is very rich in plagioclase, and geologists believe that it formed by the shattering and mixing of plagioclase-rich highland rocks. At the Apollo 16 site near the crater Descartes, similar breccias were collected, many of which contain sizable fragments of intensely crushed plagioclase-rich rocks.

Many of the breccias in both the Fra Mauro and Cayley Formations contain a great deal of formerly molten material resembling lava. Geologists believe that, at the time of the great impact events that formed the mare basins, much of the original rock was melted by the force of the impact. This molten material was included with the broken rock deposited in layers around the basins, and in many places the heat and the molten rock actually welded parts of the breccia together to form an unusually solid rock (photo 21). Many specimens of this impact melt, as such molten rocks are called, were collected from the Apollo 16 site. They resemble volcanic lava except for the large number of rock fragments they contain.

The action of heat during formation of these breccia layers has produced some unusual effects in specimens of the Fra Mauro Formation collected from the Apollo 14 site. Clustered in small cracks and crevices in the breccia samples are tiny crystals of metallic iron and a calcium phosphate mineral called apatite (photo 22). The crystals are too fragile to have grown from a liquid, and the geologists who have studied them are convinced that they were deposited directly from a heated vapor that filled the crevices soon after the layer of breccia was deposited as a mass of hot rubble on the surface of the moon. The presence of heat and gases also helped to weld the layer of originally loose breccia into a fairly solid rock.

The growth of crystals from hot gases is commonly seen around terrestrial volcanoes, where solid sulfur is the most common mineral observed. Although no sulfur crystals were detected in the Apollo 14 breccias, the minerals found indicate that the breccias were deposited on the moon at high temperatures, perhaps above 500° C. (932° F.) and that gas was present when they formed. In view of the lack of atmosphere on the moon and the dryness of lunar rocks, the evidence that a gas of unknown composition had been present when the breccias formed was one of the most provocative results from the Apollo 14 samples.

The study of lunar breccias has not been as easy, as rapid, or as straightforward as the study of the mare lavas. The lavas are a uniform and easily interpreted rock type, while a sample of highland breccia must be scanned carefully, millimeter by millimeter, to identify all the fragments in it and to decipher all the different individual histories that are preserved in a single sample. The examination of lunar breccias is not complete; on many samples, it has not even begun.

Nevertheless, the time spent on lunar breccias is well invested. In many ways, the breccias are more important to scientists than are the lunar lavas. The lavas are a single rock type with a definite age and a simple history. In the breccias scientists can find many different fragments, each with its own history, which have probably been brought together from original locations scattered across the whole moon. In lunar breccias, the scientists can observe a much wider range of rock types than could be collected from the bedrock at any one point on the moon. The breccias record events in the moon's history that took place after the solid rocks were formed. They have recorded what happens when an asteroid hits the moon, how often this has happened, how rapidly rocks are destroyed on the lunar surface, how material is transported across the moon, and what the lunar surface was like millions, or billions, of years ago.

It is not surprising that the most puzzling rock returned by the Apollo missions is a breccia—a small lemon-size rock (sample 12013) picked up by Charles Conrad from the surface of Oceanus Procellarum near the end of the first sampling excursion on Apollo 12. Analyzed on earth, its composition and age turned out to be unique, unlike that of any other known moon rock. It contains 61 percent SiO_2, whereas the associated lavas have only 35 to 40 percent. More surprising, it contains about 40 times as much potassium, uranium, and thorium as the lavas, making it one of the most radioactive rocks ever collected from the moon. Its measured age was 4 billion years, compared to the 3.3-billion-year age of the lavas from the same site.

Despite the intense effort expended upon this single rock, its origin and history are still not clear. The rock is actually composed of three different materials: a dark gray breccia, a light gray breccia, and what looks like a vein of solidified lava. How the different components of this rock formed, and what their relative ages are, is still not understood. This single rock, with its unique chemistry and its unknown original location, is a constant reminder that we still have a lot to learn about the moon.

CHAPTER

THE LUNAR SOIL

Except in the walls of Hadley Rille, the Apollo astronauts found no solid bedrock exposed on the moon. Almost everywhere the lunar bedrock is buried under a layer of loose fragmental material, usually several meters thick, that forms the actual surface of the moon. All the rocks collected by the astronauts were loose fragments and blocks from this surface layer.

The nature and origin of this layer of lunar rubble, commonly called the "lunar soil," is an important problem in the study of the moon, and collecting specimens of the soil was a major goal of the Apollo missions. Actually, it proved impossible not to collect it, for it stuck to everything. It returned to earth in sample bags and as coatings on larger rocks. It clung to spacesuits and was tracked from the lunar surface into the Lunar Module. It coated films and cameras and sent several dismayed technicians into an unexpected quarantine with the astronauts when the packages were opened. To the astronauts, the lunar soil was an acknowledged nuisance to good housekeeping in space. To the engineers, it was a potential danger to the delicate and critical machinery in the spacecraft. To geologists, the lunar soil was one of the most complex and exciting materials brought back from the moon.

The lunar soil is in fact a loose breccia containing many different ingredients. Even in the smallest fragments, as small as a fraction of a millimeter, there is tremendous variety. From a small pinch of lunar soil, geologists can separate tiny rock fragments, bits of single crystals, irregular, slaggy, glass-rich particles, and spherules and droplets of solid glass that range in color from clear or pale green to orange and red-brown.

Nearly all the rock fragments seem to come from the bedrock immediately beneath the soil. On the maria, the rock fragments are basalt, and the individual crystal fragments are pyroxene, plagioclase, ilmenite, and other minerals found in the basalt. Soils from the highlands, where they occur on top of thick layers of breccia, are composed chiefly of fragments of plagioclase-rich highland rocks and broken crystals of plagioclase.

The lunar rocks could be easily compared with terrestrial rocks, but the lunar soil was something entirely outside our terrestrial experience. Unlike terrestrial soils, lunar soil contains no evidence of the action of wind or water. It contains no weathered rocks, no organic matter, and no life. There was no word in the vocabulary of terrestrial geology to describe it accurately. Some scientists called it the regolith; most called it lunar soil and let it go at that.*

* Strictly speaking, "soil" is incorrect for the lunar material because the term describes terrestrial material formed by the weathering of rock and the activity of organisms; but neither process occurs on the moon. "Regolith" is a little more precise, because the term, as originally defined for terrestrial geology, designates a loose blanket-like deposit that forms the surface and overlies solid bedrock underneath. However, "lunar soil" is a more convenient term, and most workers in the field find it adequate.

The unique nature of the lunar soil is explained by its unique origin. The soil contains ample evidence that it was formed by the continuous impacts of meteorites—a process which no terrestrial soil or volcanic ash bed could ever have experienced.

Scientists recognized even before the Apollo Program that the surface of the moon, unprotected by an atmosphere, would be continually bombarded by large and small pieces of solid matter (i.e., meteorites), that travel through the solar system. These impacts would break up any exposed bedrock and would, over millions or billions of years, build up a layer of loose rubble. The Ranger and Orbiter pictures documented the existence of this layer of rubble in such detail that it was possible to measure the thickness of the layer (generally 5 to 15 meters) at various places on the moon.

Lunar soil samples returned by Apollo 11 offered conclusive evidence that the soil had been produced by the pulverizing effects of meteorite bombardment and not by some other process like volcanic eruptions. Evidence for the violent effects of meteorite impact is everywhere in the soil. The material that composes it has been ground very fine, and the average particle size is less than a tenth of a millimeter. All the rock fragments in the soil are angular and fractured, indicating that they were produced by the breaking of solid rock.

The solid glass droplets in the lunar soil do not resemble the jagged, irregular glassy particles found in terrestrial volcanic ash. Instead, the lunar glasses seem to have formed from jets of molten rock that were sprayed out of impact craters and then broken up into tiny beads (photo 23). Other glass fragments in the soil show the effects of the sudden, high-temperature melting produced by a meteorite impact; the partly fused crystals and incompletely mixed swirls of glasses found in these specimens are totally different from the textures produced by the orderly cooling and crystallization of volcanic rock. Many individual fragments in the soil have tiny pits, called microcraters, on their surfaces. These pits, a fraction of a millimeter in diameter and often lined with glass, are formed by the impact of dust-size cosmic particles traveling at several kilometers a second.

The lunar soil also contains remnants of the actual particles that struck the moon. Small fragments of the nickel–iron alloy of which meteorites are made have been found in samples of the soil. Tests have been made of the lunar soil for chemical elements, such as gold, platinum, and iridium, that are rare in the lunar rocks but more common in meteorites. The analyses indicate that 1 to 2 percent of meteoritic material has been mixed with the ground-up lunar rocks.

The accumulated evidence indicates that the lunar soil was formed by the continuous bombardment of the moon by cosmic particles over billions of years. No exact calculations are needed to realize what a slow process this is; it has taken 3½ billion years to build up a rubble layer less than 20 meters thick on the lunar maria.

METEORITE IMPACT: THE COSMIC PULVERIZER

Although virtually all the solid matter in the solar system is concentrated in the sun, the planets, and their satellites, there is still a tiny fraction that exists as much smaller solid bodies. These smaller bodies occur in all sizes, from microscopic bits of cosmic dust less than a thousandth of a millimeter in diameter to massive asteroids hundreds of kilometers in size. These solid particles, which cover a wide range of sizes, are given a variety of names. The smallest particles, with diameters of less than a millimeter, are called *cosmic dust* or *micrometeorites*. Larger bodies, which may have diameters of a few centimeters to a few meters, are technically called *meteoroids* when they travel through space. When one of these bodies strikes the earth's atmosphere, it becomes a *meteor*, producing a bright trail of light often called a "shooting star." If such a body passes through the atmosphere and strikes the earth, the surviving part is called a *meteorite*. Still larger bodies, hundreds of meters to hundreds of kilometers in diameter, are *planetoids* or *asteroids*. Most of these bodies are found in the *Asteroid Belt* between Mars and Jupiter, but some of them may cross earth's orbit and come quite close to it. Ceres, the largest known asteroid, is about 775 kilometers in diameter, or about one-fifth the diameter of the moon.

All this solid material travels with the planets around the sun, but the orbits of the smaller bodies often cross those of the larger planets, and occasionally one of these small bodies strikes the earth or the moon.

These solid bodies come from several sources in the solar system. Some of them, especially the larger asteroids, are "leftover" material that was not collected into larger planets when the solar system formed. However, many are of more recent origin, being produced by continuous collisions of asteroids in the belt between Mars and Jupiter. Other solid particles, especially small ones, may be produced by the disintegration of comets as they swing in toward the sun from far outside the solar system.

Before the Apollo Program, our knowledge about these smaller bits of matter in the solar system was limited to what we could learn from earth. Most of our information is relatively recent. The first recognized asteroid, Ceres, was only discovered in 1801. Meteorites were not generally accepted as genuine extraterrestrial objects until the early part of the nineteenth century, and detailed studies of fallen meteorites have been actively pursued only within the last few decades. Information about the nature and amount of dust and micrometeorites that strike the earth is still scanty.

During the 1960s, when interest in space exploration was on the rise, meteorites and other extraterrestrial matter were intensely studied for several reasons. For one, they represented a sample of the material out of which the solar system had formed. In addition, the frequency with which they hit the earth could be used to calculate the number and size of such objects in the solar system. Finally, because meteorites must hit all the planets, the effects produced when meteorites hit the earth were important for understanding how the surfaces of other planets might have formed and evolved.

Scientists learned that there was a simple relationship between the size of these solid objects and their abundance: small particles were far more numerous than large ones. Put another way, there are probably less than 20 asteroids measuring more than 100 kilometers in diameter, but millions of micrometeorite particles, which are less than a millimeter in diameter, strike the earth's atmosphere every day.

Despite the relative abundance of these smaller particles, "empty" space is still virtually empty. Only one particle as large as a pinhead is found, on the average, in several cubic kilometers of space, and both calculations and experience have shown that these particles do not pose a threat to spacecraft or to astronauts on the moon. The earth, however, is a large object, and as it swings around the sun, it sweeps up about 50,000 tons of cosmic dust in a single year, or about 135 tons a day.* Almost all of this mass enters the

* Widely varying estimates of the rate at which extraterrestrial matter falls to earth have been made by various methods. The data from micrometeorite detectors placed on earth-orbiting satellites have yielded values of from 1 to 5 million tons per year, but much of these data are now regarded with some suspicion because of the possibility of false signals produced by the detectors themselves. The lower value quoted here comes from studies of ice cores and deep-sea sediments, which accumulate slowly and thus provide a longer period of sampling time. The true value is probably somewhere between 50,000 and 1 million tons per year. This is an insignificant amount compared to the weight of the earth; it would take 4½ billion years to accumulate enough cosmic dust to change the earth's weight by one one-millionth of its value.

earth's atmosphere as particles smaller than a millimeter in diameter. Such tiny particles are not seen as they fall to earth but they have been collected from deep-sea sediments and layers in the Greenland and Antarctic ice caps. Some extraterrestrial particles have also been collected directly by high-flying aircraft. It may be difficult to distinguish genuine extraterrestrial particles from similar particles produced by industrial activity or by terrestrial volcanic eruptions. These particles are often the same size and shape as extraterrestrial ones, and careful microscopic examination or chemical tests may be needed to tell them apart.

Larger particles (meteoroids), from pinhead- to golfball-size, make known their entrance into the earth's atmosphere at once. Traveling at between 10 to 20 kilometers per second, they strike the atmosphere and burn up, causing the familiar "shooting star" phenomena seen in the night sky. These small bodies are common enough that an observer may see several on a clear night, and at certain times of the year large meteor showers like the Perseids or Leonids fill the sky with streaks of light. These displays emphasize the protection provided by our atmosphere; without it, we would be continually riddled with small particles that travel faster than bullets.

Less frequently, perhaps once or twice a year, a body weighing a few hundred grams or more will survive its passage through the atmosphere, burning off its outer surface as it slows down and strikes the ground at a fraction of its original speed. The fall of such a meteorite is often accompanied by flashes of light, a smoky trail in the sky, and loud, thunderous noises. Normally the meteorite makes a small hole in the ground, from which it is recovered to become an object for scientific study.

Although most recovered meteorites weigh less than a hundred kilograms and make only small pits in the ground, larger bodies have also struck the earth, producing far more violent effects. On February 12, 1947, a shower of large iron meteorites struck the Sikhote-Alin region of Russian Siberia, forming over thirty separate craters, the largest of which was 26.5 meters in diameter. The scientists who studied the fall concluded that the original meteorite, which weighed about 70 tons, broke into smaller pieces as it passed through the atmosphere, so that all the fragments were slowed down and struck the ground at velocities of only a few hundred meters per second. If the original mass had remained intact, it would have been only slowed down slightly by the atmosphere, and would have struck the ground while traveling at nearly 10 kilometers per second. The impact would have released the energy of about 500 tons of TNT, excavating a crater about 100 meters in diameter.

A much larger and less understood extraterrestrial object struck Siberia nearly 40 years earlier. On June 30, 1908, the region north of Lake Baikal and the city of Irkutsk, near the Podkammenaya Tunguska River, was rocked by a massive detonation as a large object passed through the atmosphere. This "Tunguska Event" was probably the most violent collision of the earth with an extraterrestrial object in recorded history:

> Ancient trees, the mighty Yenisei taiga, were torn up from their roots and in places piled up in a thick layer by the explosion wave. Hot gases scorched the surface vegetation for tens of kilometres around, and the sound of the explosion reverberated thousands of kilometres away. Seismograph observations registered an earthquake; the explosion air-wave recorded in many meteorological stations went twice round the whole world. The fall of the meteorite was seen by inhabitants at many points over a radius of 600–1000 kilometres in Central Siberia. The next few nights were phenomenally bright and shining clouds were seen in central latitudes up to 45° over Western Asia and Europe. The Earth's atmosphere, which was impreg-

nated with a large quantity of meteoric dust, was noticeably dimmed.*

Despite all the atmospheric violence, no trace of the extraterrestrial body has been found. Meticulous study of the site has yielded a few tiny glass spherules, but the nature of the Tunguska object is still a mystery. A plausible explanation is that the body was a fragile comet head, composed of small solid particles and frozen gas, that disintegrated entirely in the atmosphere. More speculative and less likely explanations are that the body was a runaway spaceship, a fragment of antimatter, or (more recently) a black hole. Whatever its nature, the Tunguska Event released as much energy as a large hydrogen bomb (about 2 million tons of TNT). Fortunately, the explosion occurred in an uninhabited area; if the object had entered the earth's atmosphere five hours later, it would have "detonated" over the city of Leningrad.

The historical record shows clearly that, within less than a century, the earth has been struck by many extraterrestrial objects weighing from a few kilograms up to more than 50 tons, and there is every reason to believe that even larger objects have hit the earth in the geological past. The effects of collisions with large objects would be far more catastrophic because large objects are not slowed down by the atmosphere. A body weighing more than about 1,000 tons—corresponding to a spherical meteorite only 7 meters (23 feet) in diameter, or about the size of a one-car garage—will pass through the atmosphere and strike the earth's surface at its original cosmic velocity, 10 to 20 kilometers per second or more. The energy released by this collision is

* E. L. Krinov, *Giant Meteorites*, translated by J. S. Romankiewicz (New York: Pergamon Press, 1966), pp. 125–26. This book gives a detailed and readable account of the scientific study of the Tunguska Event and the Sikhote-Alin meteorite shower. Both regions were so isolated that the difficulty involved just in reaching them makes a fascinating adventure story in itself.

difficult to imagine.* The impact would equal the explosion of 20,000 tons of TNT (about the energy of the first atomic bomb) and would produce a crater more than 200 meters in diameter—large enough to contain several football fields.

The impact of a large meteorite, traveling at its original cosmic velocity, is almost literally a collision between an irresistible force and an immovable object. The earth stops the speeding meteorite dead in a fraction of a second, and the tremendous energy of the meteorite's motion is transmitted into the earth as intense shock waves whose pressures may reach several million atmospheres (*Figure F*). Most of this energy is used up near the impact site to shatter and melt rock and to excavate a large crater, but a fraction of the energy can be transmitted for long distances as a seismic wave similar to those generated by earthquakes.

Scientists have developed mathematical expressions for the relations between the size of the impacting body, the energy of the impact, and the size of the resulting crater. One problem in these calculations is that the energy of impact, which equals the meteorite's energy of motion or kinetic energy (KE) depends on both the mass (M) of the meteorite and the impact velocity (V) such that $KE = \frac{1}{2}MV^2$. Thus, the same size crater can be produced by either a small body moving rapidly or by a larger body moving more slowly.

* But it is easy to calculate. The kinetic energy (energy of motion) of the meteorite is given by:

$$KE = \frac{1}{2}MV^2$$

when *M* is the mass, and *V* is the velocity at the time of impact. Substituting the mass in grams and the velocity in centimeters per second gives the energy in ergs. If we assume an average velocity of 15 kilometers/second for a 1,000-ton meteorite, then: $M = 10^9$ grams and $V = 15 \times 10^5$ cm/sec. (A number 10^n is equivalent to a 1 with *n* zeroes after it; this is a convenient way of writing large numbers.) Then:

$$KE = \frac{1}{2}(10^9)(15 \times 10^5)^2 = 11.2 \times 10^{20} \text{ erg.}$$

A kiloton of TNT (1,000 tons) is equivalent to 4.19×10^{19} erg, so the calculated energy is 26.8 kilotons.

Relations between the energy of impact and the size of the crater produced are more complicated and become increasingly uncertain for craters larger than a kilometer or so in diameter. As a general rule, the crater produced has a diameter about 50 times that of the impacting body, and the volume of rock that is shattered and excavated from the crater is at least several hundred times the volume of the impacting body. Most of this shattered and ejected rock is deposited within a distance of about two crater diameters from the crater, but a small fraction is ejected to much greater distances, especially on an airless planet like the moon. Despite the uncertainties in the calculations, the ability of large meteorite impacts to pulverize and reshape the surface of a planet is unquestionable.

We do not have to look far into the past to find the traces of such large impacts on the earth. More than a dozen impact craters are known to have formed within the last 50,000 years. They are found in desert regions where they have been preserved, and pieces of iron meteorites and fused rock are found in and around the craters. Although most of the craters are less than a few hundred meters in diameter, Meteor Crater in northern Arizona (photo 24), is over a kilometer in diameter and nearly 200 meters deep. This crater is surrounded by fragments of iron meteorites. The original body that made the crater was probably about 25 meters in diameter and weighed about 65,000 tons. The impact released energy equivalent to nearly 2 million tons of TNT—about the same amount released in the Tunguska Event of 1908.

The geological study of the earth has gradually provided evidence that even larger bodies, some of them several kilometers in diameter, hit the earth millions of years ago. These older structures have been difficult to identify. They are so old that none of the original meteoritic material is preserved, and they have been so deeply eroded that their original crater form has been erased—all that remains is a circular area of strongly deformed and shattered rock. Many geologists argued that these curious "cryptoexplosion" structures were produced by volcanic eruptions, while others felt they were caused by impacts. For several decades, the same impact–volcanic controversy that had characterized the study of lunar craters was focused on these terrestrial structures, with little resolution.

A breakthrough in the study of these structures occurred in the 1960s, when geologists developed new methods for identifying old meteorite impact structures. These new criteria are based on the realization that the tremendous energy of the impact is transmitted into the surrounding rocks by intense shock waves. As a result, the rocks around the impact crater are subjected to pressures and temperatures much higher than those developed in any volcanic eruption. These unique conditions produce distinctive and permanent changes, called "shock-deformation effects," in the rocks. These effects include intense shattering and crushing, unique deformation features in individual minerals, unusual high-temperature melting effects, and the formation

---

Figure F / The Cosmic Excavator. This series of cross sections shows the development and excavation of a meteorite impact crater produced by a body striking a series of layered rock units at 10 to 20 kilometers per second. The numbers beside the sections indicate time since the instant of impact. Initially (A) the meteorite penetrates several times its own diameter into the target rock, melting the rock near it and causing intense shock waves to radiate outward. An instant later (B) the shock waves begin to shatter the surrounding rock and eject it upward and outward. Continuation of this process (C) excavates a crater many times larger than the original meteorite. The ejected material is immediately deposited both in and around the crater as breccia (D). The final structure, formed only a minute or so after the initial impact, shows many features typical of meteorite impact craters: highly deformed and uplifted rocks in the rim, shattered and shocked rocks under the crater floor, and layers of breccia and molten impact melt in the crater itself. (Modified from E. M. Shoemaker, "Impact mechanics at Meteor Crater, Arizona," in *The Moon, Meteorites, and Comets,* B. M. Middlehurst and G. P. Kuiper, eds. (Chicago: University of Chicage Press, 1963), pp. 301–36.

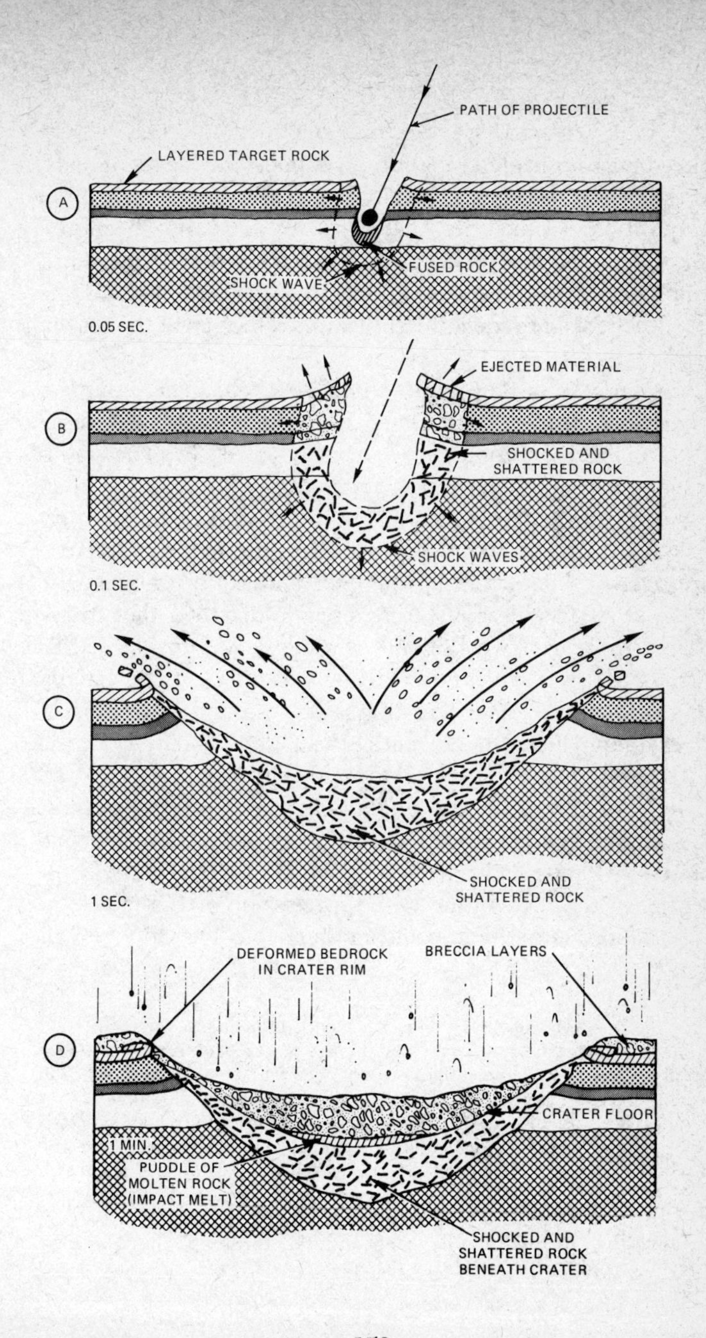
PATH OF PROJECTILE
LAYERED TARGET ROCK
A
FUSED ROCK
SHOCK WAVE
0.05 SEC.
EJECTED MATERIAL
B
SHOCKED AND
SHATTERED ROCK
SHOCK WAVES
0.1 SEC.
C
SHOCKED AND
SHATTERED ROCK
1 SEC.
DEFORMED BEDROCK
IN CRATER RIM
BRECCIA LAYERS
D
CRATER FLOOR
1 MIN.
PUDDLE OF
MOLTEN ROCK
(IMPACT MELT)
SHOCKED AND
SHATTERED ROCK
BENEATH CRATER

of strange, glass-rich breccias (photo 25). These effects have been duplicated in laboratory shock-wave experiments, in nuclear explosions, and in the rocks from young and unquestioned meteorite craters. They have not been observed in volcanic rocks, even those formed by explosions, and there is now little reason to doubt that these unique shock effects are definite indicators of ancient meteorite impacts.

In the last decade, more than 50 such ancient impact structures (called "astroblemes") have been identified by finding shock-deformation effects in their rocks. The smallest are a few kilometers in diameter, slightly larger than Meteor Crater. The largest of these structures, the Sudbury Basin in Canada and the Vredefort Ring in South Africa, are over 50 kilometers in diameter and were produced by asteroid-size bodies a few kilometers in diameter that struck the earth nearly two billion years ago.* A collision of this size is a major catastrophe by any standard. The energy released in such an event is about a million megatons of TNT—equivalent to the entire heat flow released by the earth during a 100-year period; and this energy is delivered within a few seconds at a single point on the earth's surface.

The knowledge gained from studying terrestrial meteorite craters in the decade before the Apollo 11 landing was essential to the recognition and appreciation of the effects of continuous meteorite impacts on the moon. The discovery of

* Another way of illustrating the high probability of such giant impacts in the past is to note that many asteroids of respectable size have come quite close to the earth (astronomically speaking) in recent times. In 1937 Hermes came to within 800,000 kilometers, or about twice the distance to the moon. In June, 1968, Icarus passed within 6 million kilometers, moving to pass by the sun inside the orbit of Mercury. Over long periods of time, planets like the earth tend to attract and sweep up smaller objects whose orbits lie close to theirs. It is possible that, in another few million years (a short time, geologically), Icarus will be gone and the earth will have a fresh impact crater 50 kilometers in diameter.

the shock-deformation effects in terrestrial rocks made it possible to identify similar effects in the lunar soil and to establish immediately the role of meteorite impacts in forming it. If the Apollo 11 landing had taken place only a decade earlier we would not have had this information, and the origin of the lunar soil would have been much more difficult to comprehend.

These terrestrial meteorite craters were also used as training sites for the astronauts. To cover a representative range of craters, various Apollo crews visited the small, young Meteor Crater in Arizona, the larger, older, but well-preserved Ries Kessel in Germany (diameter, 25 kilometers; age, 15 million years), and finally the ancient, deeply eroded Sudbury Basin in Canada.

In a way, Sudbury constituted my own small personal contribution to the Apollo Program. The basin, about 60 kilometers long by 30 kilometers wide, is located north of the city of Sudbury, Ontario, in a major nickel-mining region. Geological studies conducted in connection with mining activities had discovered large areas of shattered and deformed rock around the area and in 1964 geologist Robert S. Dietz suggested that the structure and the shattered rocks could be the result of a giant meteorite impact. Skeptical but curious, I made my first trip to Sudbury in 1966 and found definite shock-deformation effects preserved in the rocks. As a by-product of establishing the impact origin of Sudbury, the site was selected for astronaut training, and in 1971 and 1972 I participated in field exercises with the Apollo 16 and Apollo 17 crews. The choice of Sudbury as a training site was especially fortunate for the Apollo 16 crew, who saw many similar impact-produced breccias during their exploration at Descartes.

Meteorite impacts are especially helpful to geologists studying the moon because a single impact can scatter rock fragments across a wide area. A geologist can thus collect from one locality a variety of rock fragments derived from different places on the moon, and a single sample of lunar soil provides more information than could be obtained by sampling exposed bedrock at the same place.

Furthermore, a meteorite impact, especially a small one, is a fairly orderly process which can be duplicated in the laboratory and whose effects can be mathematically calculated. Scientists can estimate how far rock fragments will be thrown out of a given crater and from these calculations they can construct models of how the lunar soil is formed by continuous impacts.

Both the calculations and the laboratory analysis of actual lunar soil indicate that over 90 percent of the rock fragments in any given soil sample have been thrown out of small craters in the immediate area, so that the soil represents a good sample of the local bedrock even if the bedrock itself is completely covered. What is more interesting is that the same calculations indicate that many of the other 10 percent of rock fragments have been thrown out of larger craters tens to hundreds of kilometers away. The soil at any one spot contains a small number of rock fragments from far away, some of which might literally have come from the opposite side of the moon. For instance, small fragments of feldspar-rich rocks from the lunar highlands were found in the Apollo 11 mare soils nearly three years before the first landing at Descartes by Apollo 16.

In addition to scattering rocks laterally across the lunar surface, a meteorite impact penetrates into the moon and brings deeply buried material to the surface. On the average, the depth of a meteorite crater is about one-tenth of its diameter. A crater only 200 meters in diameter can penetrate through the 20-meter thickness of lunar soil and throw blocks of fresh bedrock out onto the surface around it. A giant impact like the ones that formed the large basins of Mare Orientale and Mare Imbrium may have excavated material from as deep as 50 kilometers. Careful sampling around the sites of such impacts is the only way that astronauts could obtain samples that came from deep within the moon.

The Apollo landing sites and the astronauts's lunar excursions were carefully planned to take advantage of the benefits to geological sampling provided by meteorite impacts. Both large and small impact craters played an important part in the choice of the Apollo 14 landing site at Fra Mauro. The site was selected in the hope of sampling material that had been ejected from deep within the moon by the giant impact that formed Mare Imbrium. At the landing site, the astronauts concentrated on exploring Cone Crater, a smaller crater 340 meters in diameter that had drilled through the thin layer of lunar soil and thrown out fresh blocks of the underlying breccias of the Fra Mauro Formation.

Other Apollo expeditions visited small craters a few hundred meters across to collect samples of the bedrock beneath the lunar soil. South Ray Crater provided the Apollo 16 astronauts with plagioclase-rich highland breccias, and Shorty Crater uncovered the Orange Soil for the Apollo 17 crew.

No matter where the astronauts sampled it, the lunar soil always contained a few exotic rock fragments that gave a clue to what the moon might be like hundreds of kilometers away or tens of kilometers deep. We learned much more about the moon because it had been battered for eons by meteorite impacts than we could ever have discovered if the whole lunar surface had been fresh, solid rock without a speck of dust on it.

GARDENING IN THE LUNAR SOIL

Lunar soil begins to form as soon as any fresh rock, such as a new lava flow, is first exposed on the surface of the moon

(*Figure G*). The fresh bedrock is immediately struck by cosmic particles which blast small craters in the surface and break the original rock into smaller and smaller fragments.

The continuous range of sizes of lunar craters corresponds to the varying sizes of the impacting particles. The smallest craters, a fraction of a millimeter across, are formed on the surface of exposed rocks by tiny particles of cosmic dust and micrometeorites. Large craters, from centimeters to meters in diameter, are formed by less common particles the size of marbles, golf balls, or grapefruit. The largest craters, from a few kilometers to hundreds of kilometers in diameter, record the extremely rare impacts of large meteorites and asteroids. This continuity in crater diameters, from less than a millimeter to tens and hundreds of kilometers, is one of the most convincing arguments for the existence of continuous meteorite impact on the moon.

All the impacts, large and small, contribute to formation of the lunar soil. The tiny micrometeorites shatter small rocks and dig pits in the surfaces of larger ones, while the

---

Figure G / The Making of the Lunar Soil. This series of cross sections shows the progressive development of the lunar soil layer, beginning with the moment when fresh bedrock (oblique ruling) is exposed to meteorite bombardment (A). At this early stage (B), both large and small craters (1–5) excavate bedrock and pile up layers of excavated material (open circles) in and around the craters. (Arrows indicate the direction of movement of ejected material up and out of the crater.) As these ejecta layers develop (C), large craters (6) still penetrate through to bedrock, but smaller ones (7,8) do not penetrate the rubble layer and eject only reworked lunar soil (fine dots). As the soil layer increases in thickness (D), even deep craters (9–12) fail to reach bedrock, and (E) only an unusually large crater (13) penetrates deeply enough to produce a layer of freshly ejected bedrock (fine stippling). This final stage shows some of the complicated character of the lunar soil layer, which is actually made up of many different and discontinuous ejecta layers, some of which contain fresh bedrock and others which contain only reworked lunar soil. Note also the destruction of smaller craters by larger ones during the process of soil formation. Of the 13 craters, only three (5,6,13) are still visible at the surface. Five have been destroyed (1,7,8,11,12), and the other five have been filled in (2,3,4,9,10).

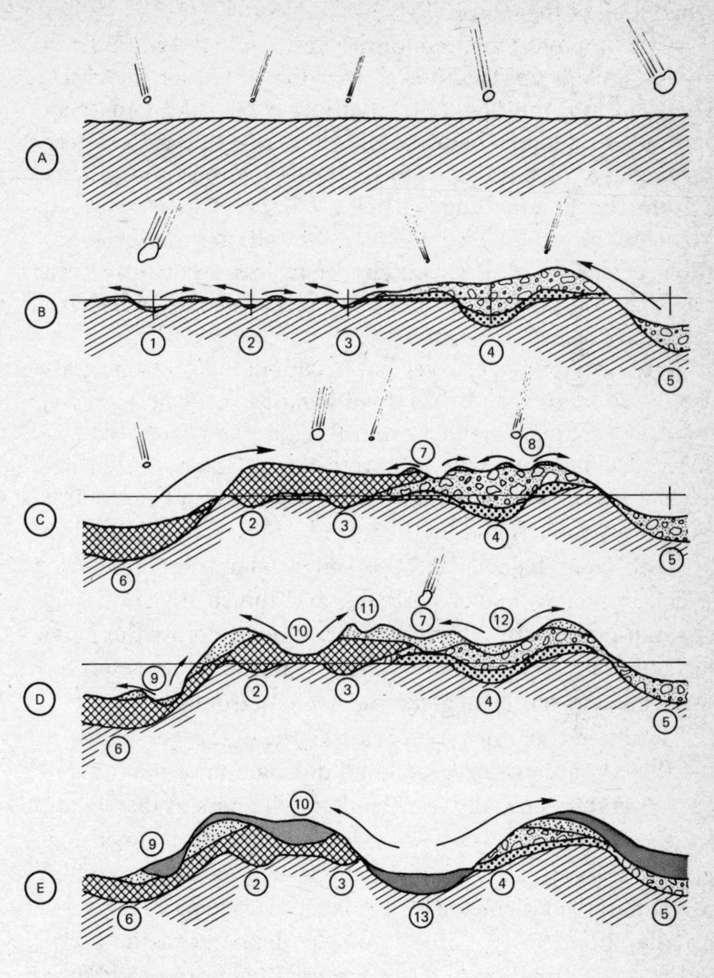
A
B
1
2
3
4
5
C
6
7
8
D
9
10
11
12
E
13
VISIBLE – 5, 6, 13
DESTROYED – 1, 7, 8, 11, 12
BURIED – 2, 3, 4, 9, 10

impact of an asteroid can scatter broken rubble over a broad area of the moon.

As the bombardment continues and more and more fresh rock is broken up, the thickness of the soil layer increases. The soil forms rapidly at first, because even small impacting particles strike fresh bedrock and shatter it. But as time goes on, the layer of soil begins to cover the bedrock and protect it from the bombarding particles (*Figure G*). Now small particles can no longer penetrate the soil, and only the impacts of larger bodies, which occur less frequently, can break through the soil to blast out pieces of the underlying bedrock.

By the time the soil layer has accumulated to a thickness of 5 to 20 meters, as it has done almost everywhere on the moon, only a giant impact excavating a crater more than 50 meters in diameter will penetrate the soil layer. Such impacts are very uncommon, and the history of a thick soil is only rarely interrupted by the introduction of freshly broken bedrock from below. But the soil is almost continuously struck by much smaller particles, and the steady formation of small craters stirs the soil and scatters it across the lunar surface. This reworking of already formed lunar soil by small impacts has been given the name "gardening."

"Gardening" on the moon is a complicated process involving the continuous formation and destruction of many small impact craters (usually less than a few meters in diameter). Craters of this size form entirely within the soil layer, and the soil excavated from them is deposited as a layer around the crater. The diameter of this layer, or "ejecta blanket," is usually about two to three times the diameter of the crater itself; only a small amount of material is ejected to greater distances. During the excavation process, large masses of soil may be removed from the crater and overturned, so that they are deposited on the lunar surface upside down. After a small crater has formed, it will be slowly buried and filled by new layers of lunar soil thrown out of younger craters which form around it.

The existence of this process of lunar gardening could be predicted from our knowledge of small meteorite bombardment on the moon and from our observations that the lunar surface is literally saturated with small overlapping craters. But the lunar samples have supplied even further evidence that stirring and overturning occur in the lunar soil. Not surprisingly, the exposed sides of some rocks collected at the lunar surface show many tiny glass-lined pits (microcraters) that were formed by the impacts of small cosmic dust particles (photo 26). However, many rocks collected by the astronauts had identical microcraters on their *undersides* as well—proof that these rocks had been flipped over at least once while they were on the lunar surface.

There is also similar evidence for such stirring and mixing within the soil itself. Samples collected several centimeters below the surface contained mineral fragments and glass spherules also pocked by microcraters (photo 27). At one time these tiny fragments must have lain on the lunar surface long enough to be struck by dust particles before being buried in the soil.

The formation of a layer of lunar soil can be divided into two stages. Immediately after fresh bedrock is exposed, a period of rapid growth occurs during which shattering of the bedrock, rather than gardening, is the dominant process. But after the soil layer has become several meters thick, the rate of growth slows down, the protected bedrock is broken up only rarely by large impacts, and the dominant process is the gardening of the soil layer itself.

If the lunar soil increased its thickness at a constant rate, i.e., so many millimeters every million years, scientists could determine the age of the lunar surface simply by measuring the thickness of the soil layer. But the fact that the soil grows rapidly at first and then slowly makes it difficult to measure its exact age. Nevertheless, the origin of the soil by continuous impacts implies that older lunar rocks would have a thicker layer of soil developed over them. Measurements obtained from various Apollo missions have shown that this relationship is generally true. The soil layer in the highlands, measured at the Descartes site, is about 10 meters thick, whereas it is only about 5 meters thick on the younger maria. Similar variations also exist between locations in the maria. At the Apollo 11 site in Mare Tranquillitatis, where the bedrock is 3.7 billion years old, the estimated thickness of the soil layer is 3 to 6 meters. But at the Apollo 12 landing site in Oceanus Procellarum, where the rocks are younger (3.3 billion years old) the soil layer averages only about half that thickness, or 1 to 3 meters, and blocks of excavated bedrock in craters only 3 meters deep were reported by the astronauts. A much younger feature, North Ray Crater near Descartes, was sampled by the Apollo 16 astronauts. Studies of rocks ejected from the crater showed that it formed about 50 million years ago; the layer of soil on its exposed rocks is only a few centimeters thick.

Although older bedrock is covered by a thicker layer of soil, attempts to calculate the exact age of the surface on the basis of soil thicknesses alone can lead to wildly incorrect answers. For example, comparison of the soil thickness might lead one to conclude that the highlands are twice as old as the maria—which would make them older than the solar system! The soil thickness can even vary greatly at the same location. At the Apollo 15 site at Hadley Rille, where 3.3-billion-year-old rocks were found, the soil layer varied between 1 and 5 meters in thickness. At the Apollo 17 site, the soil layer was complicated by clusters of moderate-size craters and by loose surface material that may have slid into the Littrow Valley from the surrounding hills; in various places, soil thicknesses ranging from 2 to 39 meters are measured.

The variation in the amount of soil present on the lunar surface is further evidence that the formation of the lunar soil is an incredibly slow process. For example, only about 5 meters of soil have formed at the Apollo 11 landing site since the lavas solidified there 3.7 billion years ago. This works out to an average rate of 1.5 millimeters of soil every million years, making formation of the lunar soil the slowest geological process ever measured. The "slow" deposition of sediment in the deep ocean basins produces about 1 millimeter of fine mud every thousand years, and thus goes on a thousand times faster.

The internal structure of the soil layer is important because it provides information about the details of the "gardening" process. The numerous impacts of small bodies—those which form craters up to a few centimeters in diameter—tend to produce a continuous stirring and mixing of the upper few centimeters of the soil. By contrast, the less common larger impacts, even though they do not penetrate the soil layer, excavate a large amount of material and deposit it as an individual layer over a relatively large area around the crater. As time passes, larger impacts build up a series of layers of ejected material, one layer for each impact, while the more numerous smaller impacts tend to mix all the layers together. Whether the lunar soil is well mixed or layered inside depends on which process goes on faster: the destruction of old layers by small impacts, or the deposition of new layers by larger ones.

From a cross section of the lunar soil, scientists can determine whether it is well mixed or layered inside, and then calculate the rates at which both large and small particles struck the moon in the past. They can compare these calculations with present rates of impact to see if the bombardment of the moon has changed with time. In addition, if layering were still preserved in the soil, it might be possible to identify the layer associated with a specific crater and thus determine when the crater had formed and what kind of rock it had excavated.

These cross sections of the soil were collected by driving a hollow metal tube vertically into the soil. When the tube was pulled out, it contained a core of lunar soil 2 centimeters in diameter and usually 10 to 20 centimeters long.

The first cores returned by the Apollo 11 mission were 10 centimeters and 13.5 centimeters long. They seemed well mixed and homogeneous, suggesting that the mixing process in the lunar soil was effective to at least these depths. But the two Apollo 12 cores told a much different story. They had penetrated to depths of 19.3 and 41.3 centimeters, and the longer core contained at least ten recognizable layers with different colors, particle sizes, and compositions. Most of these layers were composed of material identified as reworked lunar soil, but a few contained coarser particles of bedrock, suggesting that these layers recorded larger impacts that formed a crater large enough (perhaps 25 meters in diameter) to penetrate the soil at the Apollo 12 site. One particular layer, light gray in color and rich in silica, has been identified as material ejected from the crater Copernicus, 75 kilometers in diameter and about 400 kilometers away.

Still more impressive layering was found in the long core (242 centimeters) returned from the Apollo 15 landing site at Hadley Rille. This core contained at least 42 distinct layers ranging in thickness from a few millimeters to 13 centimeters. Well-layered cores were also returned by the Apollo 16 crew from the highlands near Descartes.

The cores from the later Apollo missions demonstrate that layering is preserved in the lunar soil and that the mixing of the soil by small impacts does not occur rapidly at depths greater than a few centimeters.* This conclusion agrees

* The apparent homogeneity of the Apollo 11 core may be due to any of several causes. The core might have been collected by chance from an area that had been well mixed by a large and un-

with calculated rates of soil-mixing based on our knowledge of the rate at which small particles strike the moon. The calculations show that the top one-half millimeter of the soil would be turned over 100 times every million years and should therefore be thoroughly mixed. But the stirring rate decreases drastically as one goes even slightly deeper into the soil. By contrast, the top centimeter of soil is stirred only once every *ten* million years, and to stir the soil to depths of several centimeters or meters requires billions of years.

The slowness and shallowness of lunar erosion and soil-mixing have been dramatically demonstrated by studies of the long core sample returned by the Apollo 15 mission from Hadley Rille. Measurement of unique chemical effects produced in the core by cosmic rays indicated that the entire stack of layers in the 242-centimeter-deep core had been lying undisturbed on the lunar surface for about 500 million years. The data suggest that the layers are part of a mass of lunar soil that was ejected from a large impact crater and deposited on the lunar surface as a single unit 500 million years ago. So slow has been the alteration of the moon since then that the layers have been preserved, even though they have been lying on the lunar surface for a period of time equivalent to the entire record of fossil life on the earth.

By the end of the Apollo Program, we understood some aspects of the nature and origin of the lunar soil better than we understand the soils on the earth. For all its strangeness, the lunar soil has developed by the straightforward, continuous bombardment of the lunar surface by meteorites, and it contains none of the unpredictable products of wind,

---

recognized impact. Alternatively, the thick soil at the Apollo 11 site may have been so uniform that there were no differences in color or texture by which whatever layering was present could be recognized. Finally, the layering could have been destroyed during the collection, transport, and processing of the core itself. Regardless of the reason, the Apollo 11 core sample does not seem typical of the soil in other parts of the moon.

water, and life that complicate the soils of our own planet.

Moreover, all the soil ever formed on the moon is still there on the surface, more or less unchanged. In layers only a few meters thick, the lunar soil preserves the entire record of billions of years of slow evolution on the moon. Nowhere on earth can we find any deposits that compress so much time into such a thin layer. The deep core from Apollo 15 returned a section of lunar history that had been unchanged for half a billion years. The 5-meter thickness of lunar soil at Mare Tranquillitatis contains information from the last 3.7 billion years. And the thicker soils in the lunar highlands may record the history of the moon from even earlier times.

The lunar soil is more than just the ground-up and scattered rubble of the moon's solid rocks. It is the boundary layer between the moon and outer space, and in it the history of the moon is mingled with the history of other matter and energy in the solar system. The lunar soil records the bombardment of the moon by solid particles of all sizes; it contains traces of the passage of energetic cosmic rays; and it is saturated with atoms of the solar wind that have poured out of the sun for billions of years. Even if no life grows in the lunar soil, we will one day reap a valuable harvest from it—a wealth of knowledge about our universe.

CHAPTER

OF TIME AND THE MOON

Time and the moon have always been closely related in man's mind. The moon has been our timekeeper; its regular passages through the sky have given us the months and seasons with which we measure out our own brief lives. Yet as a timekeeper, the moon has also seemed eternal. It has been with the earth for eons before the appearance of man —there is evidence of tides caused by the moon to be found in rocks buried beneath the earth's oceans more than three billion years ago.

When men began to observe the moon more closely, they noticed that it even *looks* old. Its battered and heavily cratered surface seems to preserve primordial catastrophes of which no trace is left on the surface of the earth.

Man's relatively new ability to measure the actual ages of rocks is probably his most important tool for understanding the history of his own planet and the moon. These age-dating techniques depend on certain radioactive elements in the rocks—elements which transform slowly into new substances at a steady rate over millions, or even billions, of years.

HOW ROCKS TELL TIME

Only about half a century has passed between the discovery of radioactivity and its application as a routine method for measuring the ages of rocks and minerals. Just before 1900, Henri Becquerel (1852–1908) and Marie (1867–1934) and Pierre Curie (1859–1908) discovered that certain natural atoms, like uranium and thorium, were unstable and gradually changed into lighter atoms by emitting small atomic particles. The change from the original "parent" atom to a new, or "daughter," atom goes on at a constant rate, and each disintegration releases a small amount of heat. For many radioactive elements, this rate of change is so slow that it would take billions of years for all of the parent element to transform ("decay" is the technical term) into the daughter element.

This discovery produced an immediate and violent upheaval in our ideas about the age of the earth. At the time radioactivity was discovered, the age of the earth was the subject of an active debate between geologists and physicists. Geologists, who measured the rate at which sediments were deposited in the oceans, argued that at least several hundred million years (and perhaps one or two billion years) were needed to deposit all the thick layers of sedimentary rocks now exposed on the continents. The physicists, among whom Lord Kelvin (1824–1907) was most prominent, estimated from the rate that the earth was losing heat that the earth could not be more than a hundred million years old (and perhaps no more than 25 million), even if the earth had been completely molten when it formed.

The discovery of radioactivity in rocks settled the controversy in favor of the geologists and the old earth. For one thing, radioactivity within the earth was a significant source of heat that was understandably not included in Lord Kelvin's calculations. Now the earth could be losing heat at its present rate and still be several billion years old. More important, radioactivity provided techniques for proving that rocks several billion years old actually existed on the earth.

Soon after the discovery of radioactivity, scientists established that the element uranium decayed into lead at a fixed rate. Thus, any mineral containing uranium would gradually accumulate lead, and the amount of lead produced would depend on the amount of uranium present and the age of the mineral. In 1907, B. B. Boltwood (1870–1927) dated some ancient rocks as about 1 to 2.5 billion years old by measuring the amount of lead present in uranium-rich minerals that they contained. Later workers improved on Boltwood's methods, but his basic results remained unchanged, and gradually scientists accepted the idea that the earth had existed for billions of years.

As the result of progress in several different fields, the measurement of rock ages became more routine and more accurate in the following half-century. Physicists learned a great deal more about the processes of radioactive decay. Better and more sensitive instruments were built to make the measurements. And geologists, slowly realizing the potential of this new tool, began to search diligently and successfully for even older rocks.

By 1950, several natural radioactive elements that could be used for dating rocks had been discovered. In some cases these atoms are two natural isotopes* of the same elements,

* *Isotopes* are atoms of the same element which have different atomic weights. The chemical properties that determine an element depend on the numbers of positive atomic particles (protons) and negative particles (electrons) in the atom. Because the positive and negative charges must balance, the numbers of protons and electrons must be equal, and this number (called the *atomic number*) determines the element. But an atom can also contain neutral particles (neutrons) whose number may vary slightly. The sum of the protons

such as uranium-235 (which decays to lead-207) and uranium-238 (which decays to lead-206). Another atom, thorium-232, decays to lead-208. Other useful parents for age-dating are radioactive isotopes that occur as a small fraction of a nonradioactive element. For instance, potassium is not thought of as a radioactive element, but it contains a small amount of one isotope, potassium-40, which decays to the gas argon-40 slowly enough to be suitable for dating rocks. Rubidium-87, which decays to strontium-87, is similarly useful.

As the accuracy of age-dating methods improved in the two decades before the Apollo Program, scientists were able to establish fundamental boundaries on the age of the earth and the universe. Measurements on meteorites gave ages of about 4.6 billion years, and this figure is generally accepted for the age of the earth and the rest of the solar system. Rocks in South Africa with ages of 3.3 to 3.5 billion years were thought to be the oldest rocks preserved on earth until rocks 3.8 billion years old were found in Greenland a few years ago. In the fifty years since Lord Kelvin, the "known" age of the earth expanded a hundredfold to a length of time so great that it is difficult to imagine. If the age of the earth were thought of as a single day, the existence of virtually all fossil life would cover only the last 3 hours, and the entire history of both primitive and civilized man would be compressed into the last second.

A major problem in making age measurements is that the parent and daughter elements usually occur in amounts of a few parts per million (one part per million = 0.0001 percent) in most rocks and minerals. The radioactivity of the parent elements is not dangerous and can, in fact, be de-

---

and neutrons gives the *atomic weight*, which is used to designate different isotopes of one element. Because of their importance in nuclear reactions, the isotopes of uranium have become well known. Both isotopes have 92 protons, but one isotope contains 143 neutrons (uranium-235) while the other contains 146 (uranium-238).

tected only with sensitive instruments. Analysis of the parent and daughter elements requires complicated machines called mass spectrometers, which use powerful magnets to separate out and measure the different isotopes of the same element. Sometimes the amounts of parent and daughter elements are so low that the analyses have to be made in special "clean rooms" designed to keep out atmospheric dust and reduce contamination. Under such conditions it becomes possible to measure the isotopes in a billionth of a gram of material. Scientists have measured the ages of lunar rocks that weigh less than a few thousandths of a gram and are only a fraction of the size of a common aspirin tablet.

Although the technology for age-dating is complex, the principles involved are simple, and the same physical laws apply to radioactive decay on both the earth and the moon. When a rock forms, radioactive parent atoms are trapped in different crystals, where they begin to decay into their daughter atoms. Each crystal in the rock behaves like an hourglass in which sand falls from the top (parent) to the bottom (daughter) at a known rate. Time can be determined by seeing how much sand has piled up in the bottom of the hourglass, or by measuring how much daughter element has accumulated in the crystal.

To measure actual ages of rocks and minerals, one must first determine the rate at which a sample of the parent radioactive element decays to the daughter. The age of the rock can then be calculated from the relative amounts of parent and daughter elements that it contains. The older a rock, the more of the daughter element should be present. The amounts of the parent and daughter element, combined with the decay rate of the radioactive parent, give a value for the age of the rock.

Defining the "formation" of a rock is not as obvious as it might seem, because the radioactive clocks are strongly affected by temperature. At high temperatures (500 to 1,000° C.), the radioactive parent elements can migrate freely through a rock, and the random mixing of parent and daughter elements precludes any patterns that can be used to measure the passage of time. As far as the parent radioactive elements are concerned, "formation" is really the cooling of the rock to the point where they can no longer move freely through it and become trapped in crystals. A freshly molten lava flow, or a mass of older rock melted and heated by a meteorite impact, will not begin to measure time until it has cooled below about 500° C. Fortunately, such rocks normally cool quickly enough so that the "age of formation" recorded by the radioactive clocks is not significantly different from the time that the lava erupted or the meteorite struck.

Certain other conditions in nature have to be met if the measurements are to yield valid ages. First, the sample must have formed at a single moment in time, so that it has a meaningful age. This means that the rock should have been heated and cooled throughout in a single brief episode. "Ages" determined on a complicated mixture like a lunar breccia or a continually reworked material like lunar soil do not reflect a single moment of formation.

Another complication is introduced when the rock already contains some of the daughter element at the time that it forms; in this case, the measurements will make the rock seem older than it really is. Further problems arise if any chemical changes occurred in the rock after it formed. If some process adds more parent element or removes the daughter element, then the rock will appear too young. Conversely, if parent is removed or daughter added, the rock will appear too old. Finally, if some process heats or alters the rock slightly, so that parent and daughter elements can move from their original sites to other locations in the rock, the results may become very difficult to interpret.

All these difficulties have been encountered in trying to age-date terrestrial rocks, especially in geological environments where water is available to alter the rock or help move elements around. Geologists have developed ways of correcting for these effects and obtaining the true age of the rock in spite of them. One of the best procedures is to measure several different parent–daughter element pairs in the same rock and to calculate an independent age for each of them. If all the ages are the same, then it is likely that they all record the true time of formation of the rock.

These analyses can actually be used to measure two different kinds of ages of a rock: the *formation age*, which gives the time since the rock cooled and radioactive atoms became trapped, and a second kind of age called a *model age*. The model age is determined by total chemical patterns carried in the whole rock rather than in individual crystals. Instead of giving the time when the rock formed in its present state, a model age provides clues to earlier events; it may, for instance, indicate the age of the material out of which the rock was formed.

To illustrate the difference between model ages and formation ages, imagine taking a chunk of granite with a measured formation age of 2 billion years and grinding it into powder. The powder you have just made has a formation age of zero. But because the chemical patterns remain the same in the powder as they were in the original rock, measurements on the powder will yield the same 2-billion-year model age. If you melt the powder in a furnace, the resulting glass will also have a formation age of zero while its model age remains at 2 billion years.

Many rocks have relatively young formation ages and much older model ages; both ages provide important information about the origin and ancestry of the rock. In the case of lavas with young formation ages, an older model age may be inherited from the original rock which was melted to produce the lava. The Apollo 11 lavas, for example, have formation ages of 3.7 billion years, but their model ages are greater than 4 billion years. This suggests that lunar material more than 4 billion years old was melted to produce the lavas 3.7 billion years ago.

Complete melting of a rock, which could follow a large meteorite impact, would thoroughly mix the parent and daughter atoms and erase all the original formation ages. When the molten rock cooled, it would develop a new formation age that dated from the time of the impact. However, despite the melting, the newly formed impact melt would still preserve some information about the nature and age of the original rock struck by the meteorite.

AGES OF THE MOON ROCKS

The Maria

We were fortunate that the first samples returned from the moon were fairly simple lavas from the maria. The rocks had formed during a single period of cooling of molten lava, and microscopic examination of the rocks detected no sign of alterations that would indicate that the radioactive clocks had been disturbed by later events. The ages were easily measured. Different parent–daughter combinations (especially the rubidium–strontium and potassium–argon pairs) gave the same results. Furthermore, different specimens collected at the same site recorded the same ages. Few terrestrial rocks had ever been as easy to interpret as the lavas from Mare Tranquillitatis. The ages of 3.7 to 3.8 billion years were established as the time of eruption and cooling of the basalt lava flows.*

* Samples from a single lunar site show a range in age of about 0.1 to 0.2 billion years for locations on the maria, and the range is even wider in the highlands. Some of this scatter is due to the experimental uncertainty in each measurement. This uncertainty arises from the fact that none of the laboratory measurements can be made with absolute precision. As a result, any stated age might be in error by 3 to 5 percent, which, for a 4-billion-year-old rock, is about 0.1 to

The ages of all mare basalts are similar and fall into a narrow range between about 3.2 and 3.8 billion years. The basalts from Mare Tranquillitatis (Apollo 11) and Littrow Valley (Apollo 17) are the oldest at 3.7 to 3.8 billion years —about the same age as the oldest preserved terrestrial rocks—whereas those from Oceanus Procellarum (Apollo 12) and Hadley Rille (Apollo 15) are younger—3.2 to 3.4 billion years. The lava from Mare Fecunditatis, sampled by the unmanned Russian lander Luna 16, is intermediate in age (3.4 billion years). Thus, the pattern that emerged indicated widespread flooding of the moon by lava for about half a billion years, followed by an apparent end to volcanic activity about 3.2 billion years ago.

The painstaking analyses needed for these measurements also produced some important chemical information about the moon. The patterns of radioactive parent and daughter elements in moon rocks differ greatly from those in terrestrial rocks. In particular, the ratios of different isotopes of lead are totally dissimilar, which suggests that the moon and earth have been completely separate planets since they formed. The lead in terrestrial rocks contains an isotope (lead-204) that does not form by radioactive decay. The presence of this isotope indicates that the earth contained a significant amount of original lead when it formed. The lead in lunar rocks is totally different. It contains almost no lead-204; its lead is almost completely made up of isotopes derived from the decay of uranium and thorium.

This indicates that the moon had practically no lead when it formed. The lunar rocks also showed a similar low content of rubidium. Both rubidium and lead are volatile elements, easily lost by heating, and their absence in lunar rocks was more evidence that the moon's material had been subjected to intense heating, during which rubidium and lead, to-

---

0.2 billion years. However, some of the scatter, especially in highland rocks, probably reflects slightly different times of formation of the individual samples.

gether with more common substances like sodium, potassium, and water, were boiled away and lost, possibly even before the moon itself formed.

The Lunar Soil

Scientists made the same analyses on the lunar soil as had been made on the rocks. They did not expect to determine a formation age for the soil, because the lunar soil has a complex disparate composition that forms and mixes continuously, and its history should contain no single event to which a formation age could be assigned. Instead, the analyses of the soil were expected to produce chemical data that might explain the processes of soil formation and the ages of the different materials contained in the soil.

The results were quite unexpected. Samples of the lunar soil have model ages of 4.4 to 4.6 billion years—more than half a billion years older than the ages of the rocks they covered. No matter where the soil was collected or how many samples were analyzed, the same figure of about 4½ billion years came up.

This discovery provoked considerable questioning among the scientists involved in these analyses. If the lunar soil is formed continuously out of a variety of different materials, how can it have a unique model age, let alone the same model age in many different parts of the moon? And if the lunar soil is formed mostly by the breakup of local bedrock, how can it appear to be older than the rocks it has formed from? Finally, was it only coincidence that the model ages of the lunar soil were the same as the ages measured on meteorites, which also date the formation of the solar system? There appeared to be a serious gap in our knowledge somewhere.

What these unusual results suggested was that there had to be more in the lunar soil than just broken-up local bedrock. Some unrecognized factor was producing the high model ages. For want of a better term, scientists called it the "magic component" and started looking for it.

Careful separation and examination of the lunar samples yielded a number of unusual rock fragments mixed in with the bits of local bedrock that made up 90 percent of the soil. Some of these exotic rocks were plagioclase-rich gabbros and anorthosites from the lunar highlands; their chemistry could not explain the high model ages scientists were trying to account for. Other strange fragments were found in the soil, mostly tiny pieces of crushed and shattered rock and unusual yellow-brown glass (photo 28). These fragments were first recognized in the Apollo 12 soils, and even larger pieces of the same materials were found in the breccias brought back by the Apollo 14 astronauts from Fra Mauro.

These exotic rocks greatly resemble the mare basalts, and, like them, are composed chiefly of the minerals plagioclase and pyroxene. But they contained several times as much of the radioactive parent elements rubidium, uranium, and thorium as did either the mare basalts or any rocks collected from the lunar highlands (with the single exception of the unusual breccia 12013; see p. 139). Furthermore, the model ages determined for these small fragments were the same as those of the lunar soil itself, about 4.4 to 4.6 billion years.

The composition of these rocks was unusual in other ways. In addition to containing more uranium and thorium than other lunar rocks, they also contained greater amounts of potassium (K), a group of metals known as the rare-earth elements (REE), and phosphorus (P). These letters were combined to name these unusual rocks "KREEP material" or "KREEP basalts." (They are also known as "Fra Mauro basalts" after the Apollo 14 landing site where they were first collected in quantity.)

KREEP material cannot be too common on the moon. If it made up a large volume of the moon's interior, its high radioactivity would supply enough heat to keep the lunar interior completely molten, and the moon would be erupting lavas to this day. From what we have learned from the Apollo landings and from the analyses made from lunar orbit, it seems that most of the KREEP material occurs around Mare Imbrium and possibly in one or two other areas, where it produces the radioactive "hot spots" detected from orbit (see p. 114). (The orbital analyses revealed that the rest of the lunar surface has a much lower radioactivity that is characteristic of rocks like the mare basalts and the highland anorthosites.) The KREEP material probably occurs at depths of 25 to 100 kilometers under Mare Imbrium, and was blasted out and scattered across the moon by the impact that formed the Imbrium Basin. KREEP is common in the Fra Mauro Formation (which is composed of material thrown out of Mare Imbrium), and it is also found in soils from the Apollo 15 and Apollo 17 sites, both of which are close enough to Mare Imbrium to have been reached by the ejected material. The high radioactivity makes the KREEP material an important source of heat; it may be partly responsible for the high heat flows measured at the Apollo 15 and 17 landing sites. If there is a great deal of KREEP material under Mare Imbrium, it may have generated enough heat to play a major role in the production of lunar lavas 3½ billion years ago.

Further study of the KREEP material indicated that it is the "magic component" that makes the lunar soils seem old. At most sites, it makes up less than 10 percent of the soil, mostly as very fine material that is hard to recognize by microscopic examination. But the KREEP material is so enriched in radioactive parent elements that it dominates the soil and imposes its own model age on it, just as a small amount of black paint can turn a large amount of white paint into a dark gray.

Discovery of the KREEP material vindicated scientists' original ideas about the lunar soil. The lunar maria were not more than 3.8 billion years old and scientists were confident that the soil formed on them was even younger. But the old model ages of the KREEP material suggested that somewhere on the moon there were rocks which recorded lunar events over 4 billion years old and might even extend back to the formation of the moon. If such old rocks still existed, they would be in the lunar highlands.

The Highlands

Long before the Apollo 11 landing, telescopic mapping of the moon had established that the highlands were older than the maria. The Apollo 11 samples dated the formation of Mare Tranquillitatis at 3.7 billion years ago, indicating that the highlands must be even older. But the same telescopic studies had shown that the highlands were covered by overlapping craters tens of kilometers in diameter that recorded an intense and savage bombardment of the moon. Any original highland bedrock must have been repeatedly melted, shattered, mixed, and spread across the moon. Although the highlands contained the early history of the moon, it would not be easy to read. Scientists could only wait for actual samples to see how much of the record could be deciphered.

The first Apollo landings afforded some hints that records of older events did survive in the highlands. The small bits of white anorthosite in the mare soils furnished data on the nature of the highlands, but the rocks were too small for age measurements. However, the unusual breccia 12013, which was probably a highland rock, registered a model age of 4.0 billion years. Even older model ages, about 4.4 billion years, were obtained for the KREEP material in the lunar soils.

These early results encouraged scientists while they waited for landings in the highlands. If the ancient model ages actually reflected widespread lunar events, then rocks formed during these events might still exist in the highlands. It seemed likely that any highland rock would be older than the oldest rocks ever found on earth. And it was reasonable to hope that somewhere in the highlands the first rocks formed on the moon 4.6 billion years ago might be preserved, either in the boulders from the Apennine Mountains, in the Cayley Formation at Descartes, or on the hills around the Littrow Valley.

Highland rocks were collected by four Apollo missions (14, 15, 16, and 17) from four locations separated by hundreds of kilometers. Despite the exciting fact that many of the samples were dated at more than four billion years, none of the rocks were as old as the optimists had hoped. The ages of rocks from Fra Mauro (Apollo 14) clustered around 3.9 to 4.0 billion years—presumably the time of formation of the Imbrium Basin. The samples, mostly breccias, brought back by other missions, were only slightly older, 4.0 to 4.2 billion years; the Genesis Rock (sample 15415) from Apollo 15, for instance, was 4.1 billion years old. Regardless of the location, no highland sample had a formation age as old as the supposed formation of the moon—4.6 billion years ago.

Some highland samples do contain rare and tantalizing traces of older events. One specimen, an olivine-rich breccia (sample 72417) collected by the Apollo 17 astronauts did yield a model age of 4.6 billion years, but the rock itself is so severely shattered and deformed that its history is uncertain, and it is too early to regard it as a piece of the original lunar crust.*

In other samples collected from Descartes (Apollo 16), scientists found crystals of the mineral plagioclase that seemed to have two different ages. The outer rims of the

* A more definite age of 4.6 billion years was recently reported at the Seventh Lunar Science Conference in March, 1976. The age was measured on a crystalline rock (sample 76535) returned by the Apollo 17 mission. It now seems clear that, despite the intense early bombardment of the moon, a few lunar rocks still preserve the record of the moon's formation and earliest history. These results also mean that the Apollo 17 mission *did* succeed in its goal of finding the oldest lunar rocks.

crystals gave ages of about 4.2 billion years, the same as those measured on the rocks, but the inner cores of the crystals were apparently older, with ages of 4.4 to 4.5 billion years. A satisfactory interpretation of these results has not yet been established. One possible explanation is that the crystals actually did form 4.5 billion years ago and that their radioactive clocks were partly reset by heating 4.2 billion years ago, while also leaving the older formation age preserved in the unaltered cores of the crystals.

Although analyses of the highland rocks indicated formation ages of 4.0 to 4.2 billion years, the highland soil was found to have model ages of 4.4 to 4.6 billion years. This finding suggested that KREEP, or some similar material, has been mixed into the soils of the highlands as it had been in the soil on the maria.

Why didn't we find the really ancient rocks that we expected in the highlands? One reason may be that we were just fantastically unlucky in our sampling. Perhaps older rocks remain to be found somewhere on the moon, in distant highlands, near the poles, or perhaps on the far side. This explanation seems unlikely, however, since the four highland landing sites were carefully selected to provide a maximum chance of finding old rocks. Not only were the sites themselves separated by hundreds of kilometers, but each single site contained material thrown into it by meteorite impacts over a wide area of the moon. We can be fairly sure that the highland samples are representative of a large part of the moon and that few, if any, specimens older than 4.2 billion years will be found.

If we accept the evidence provided by meteorites and by the high model ages of the lunar soil, then the moon must have formed about 4.6 billion years ago. If we cannot find the oldest highland rocks, then something may have destroyed most of them, just as the earliest rocks formed on the earth were destroyed.

On our own planet, this absence of our early record is

easy to understand, for we can see with our own eyes the continuous destruction and re-creation of rocks by the forces of erosion, volcanism, and mountain-building. But there is no indication that such processes have been continuously active on the moon. Widespread volcanism apparently ceased on the moon more than 3 billion years ago, and the only change we now observe is the slow formation of the thin layer of lunar soil—a process too slow to be capable of destroying the whole record of the early history of the highlands.

From the data we have secured, it seems that the early highland rocks must have been destroyed between 4.4 and 4.0 billion years ago by some process that affected the whole planet. One process that must have been going on at about this time was the series of giant impacts that formed the large mare basins.

The catastrophic formation of Mare Imbrium is dated at about 3.9 billion years ago. Geological observations have shown that Imbrium is only one of the youngest of more than 15 large circular basins on the near side of the moon. The Orbiter photographs have shown that a similar number of equally large basins occurs on the far side of the moon, although the far-side basins are not filled with lava flows. It seems clear that, between the time the moon formed and about 3.9 billion years ago, the moon was struck by more than two dozen large asteroid-size bodies as well as many smaller ones. The melting, shattering, and mixing of the original highland rocks by these separate catastrophes could be the means by which the ancient rocks were destroyed and their radioactive clocks reset to the younger ages that we now find.

This catastrophic modification of the highlands during the first half-billion years of the moon's existence could have occurred in two different ways, and scientists are divided as to the details of what actually happened. One group argues that all the large bodies hit the moon suddenly in a brief "terminal catastrophe" between 3.9 and 4.0 billion years ago. This mechanism would explain the fact that so many of the lunar rocks have formation ages that fall into this interval. The older formation ages (4.0 to 4.2 billion years) would come from rocks whose radioactive clocks had not been entirely reset by the bombardment.

However, the "terminal catastrophe" idea does not explain *why* so many large bodies struck the moon as late as half a billion years after the moon had formed. If the bodies had been anywhere near the moon, they would have been swept up by it very soon after it had formed. Some special mechanism would be needed to "store" these bodies out of reach of the moon for half a billion years and then send them crashing into it all at once.

This "storage problem" is avoided in the arguments of other scientists who do not believe that the "terminal catastrophe" ever occurred. Instead of undergoing a sudden barrage of large impacts 4 billion years ago, they argue, the moon has been continually struck by both large and small bodies since its formation so that the older rocks have been continuously destroyed or altered by younger impacts. In this view, the various mare basins were formed by large impacts that occurred less than 50 to 100 million years apart. The ages estimated for the mare basins by these scientists range from about 4.5 billion years for the oldest (such as Mare Nubium and Mare Tranquillitatis) to 3.8 to 3.9 billion years for the youngest (Mare Imbrium and Mare Orientale).

At present we do not know which theory is correct. Either can explain the pattern of ages that we now observe in the highland rocks. As often happens in science, both views may be partly right. There must have been an intense bombardment associated with the formation of the moon by accretion, and a sudden swarm of large impacts might have been superimposed on the last waning stages of this original bombardment.

In any case, it is clear that the intense bombardment of the moon ceased about 4 billion years ago, because the rocks that formed after that time are generally preserved. Why the bombardment decreased so dramatically at that time is not entirely understood; one possibility is that virtually all of the available asteroid-size bodies near the moon may have been swept up by the earth and moon, leaving relatively little solid material in the immediate vicinity.

Evidence for an early intense bombardment of the moon actually predates the Apollo Program. Scientists who made estimates of the relative ages of lunar surfaces from the number and size of craters developed on them observed that the highlands were far more cratered than the maria. If the bombardment rate of the moon had been constant since the moon formed, then the highlands had to be very old and the maria very young. When the Apollo samples established that the maria were old, it became clear that the highlands must have been bombarded more intensely during the moon's first half-billion years in order to form all the craters that were there.

Crater counts show that any given area in the highlands contains about 30 times as many large craters (1 to 100 kilometers in diameter) as the same area of a mare surface. The highlands are literally saturated with large, overlapping craters in the same way that a battlefield is saturated with craters formed by an intense artillery barrage, so that older craters are being destroyed as fast as new ones are formed.

In their first billion years the highlands were bombarded 30 times more heavily than the entire moon has been during the 3½ billion years that followed. If one tries to calculate farther back toward the origin of the moon, the bombardment rate approaches infinite values and the mathematical equations break down; so it is not surprising that the oldest highland rocks are not preserved on the moon. In fact, as the details of that half-billion-year bombardment become clearer, it seems almost surprising that any rocks 4.2 billion years old have survived at all.

One by-product of this early bombardment of the highlands was the development of lunar soil on a grand scale. By the time the bombardment ended 4 billion years ago, the highlands were covered with a layer of rubble several hundred meters to a few kilometers thick, made up chiefly of separate layers of material excavated from each of the large mare basins, with smaller contributions from the millions of smaller craters that were formed at the same time. Because of this thick cover of rubble on the highlands, it is not surprising that the samples collected are all complex breccias, and it is very doubtful that any original bedrock will ever be found exposed in the highlands.

THE AGE OF THE MOON

The lunar samples, with a few exceptions, have ages of less than 4.3 billion years, and the idea of a catastrophic bombardment of the young moon has been proposed to explain the absence of older rocks. This explanation rests on our assumption that the moon did form at the same time as the meteorites, 4.6 billion years ago. If this assumption is wrong, then the moon might be younger than the rest of the solar system. Could the moon have formed about 4.2 to 4.3 billion years ago, the age of its oldest known rocks, after existing as dispersed matter for several hundred million years after the meteorites had formed? Based on evidence provided by the moon rocks themselves, the answer to this question has to be no.

If our present ideas about the origin of the solar system are correct, then the moon and the planets formed by the contraction and condensation of a huge whirling cloud of dust and gas. All during the time that the material in the original dust cloud remained well mixed and chemically homogeneous, radioactive parent atoms like uranium and rubidium were decaying steadily to daughter elements. As time passed, the amount of daughter elements increased steadily in the cloud. The longer that the cloud remained well mixed, the more daughter elements it contained, and the higher the amounts of daughter elements in any solid matter that finally separated to form meteorites, the moon, or the planets.

By measuring the amounts of daughter elements, especially strontium and lead, in meteorites and lunar rocks, scientists were able to show that the solid matter that makes up these bodies could not have stayed in a well-mixed dust cloud for more than a few million years. The exact figure is uncertain, and different chemical studies give different values from about 5 million years to about 20 million years. Nevertheless, the data are accurate enough to show that the moon could not have formed several hundred million years after the other planets. If it had, all its rocks would contain much more of the daughter elements than has been found. Furthermore, evidence that the moon existed as an individual object about 4.5 to 4.6 billion years ago is provided by the high model ages of the KREEP material in the lunar soil.

The analyses of lunar rocks therefore support the mathematical models which suggest that it should have taken no longer than a few million years to form the sun and the planets from the original dust cloud. It is therefore likely that the earth, the moon, and the other bodies of the solar system all formed together at this time. But the chemical data show that fundamental differences in chemistry already existed between the earth and the moon at the time of formation—for instance, the general depletion of the moon's rocks in such elements as hydrogen, sodium, potassium, and lead. This strongly suggests that the earth and moon were never part of the same planet.

CHAPTER

THE LUNAR SURFACE: THE BILLION-YEAR SUNBURN

The surface of the moon is a desolate and hostile place far outside of man's previous experience. Compared to the lunar surface, the earth's dry, scorching deserts and wind-swept frozen icecaps are comfortable, friendly, and swarming with life. Because the moon has no atmosphere, the matter and energy that strike its unprotected surface from the rest of the universe create an environment in which nothing can live. In direct sunlight, the surface temperature rises above the boiling point of water, only to plummet far below freezing during the lunar night. A steady rain of cosmic particles wears down exposed rocks and digs small craters in the lunar soil. And lethal, high-energy atomic particles from the sun and the stars burrow into the lunar soil, leaving permanent traces of their passage.

For more than four billion years, the moon's surface has been the scene of a continuous battle between its hard, durable rocks and the steady, destructive bombardment from space. Continuing impacts stir the lunar soil layer and add to it, burying material that lay on the surface and exposing newly broken bedrock and deeply buried soil. This slow overturning has gone on for billions of years, while the upper part of the soil has been exposed to the full fury of the sun.

The star we call "the sun" is essentially a huge nuclear reactor. It produces energy by "burning" its hydrogen atoms to form helium, releasing an amount of energy equal to the explosion of a billion hydrogen bombs every second. Even on the earth, 150 million kilometers (93 million miles) from the sun, raw sunlight would be immediately fatal to life if our atmosphere did not remove the lethal ultraviolet radiation, X-rays, gamma rays, and high-energy atomic particles. Only within the last two decades have we cautiously sent instruments, and then men, above our atmosphere to observe the true nature of the sun. Yet the unprotected lunar surface has been exposed to this gigantic nuclear reactor for billions of years.

Only a small fraction of the sun's energy that falls on the moon is permanently trapped by the lunar soil. The heat absorbed during the 14-day lunar "day" is radiated back into space again as the soil cools during the following two-week lunar "night." Primary X-rays from the sun are quickly transformed into secondary X-rays that are emitted into space again. Only the highest-energy atomic particles from the sun penetrate deeply into the lunar soil to cause permanent changes.

The effects produced in the lunar soil by the sun are similar to changes produced in small samples exposed for hours or days in man-made nuclear reactors on earth. Nuclear transformations occur, producing new kinds of atoms, and high-energy atomic particles leave permanent tracks as they pass through crystals and glass fragments. But the irradiation of the moon by the sun occurs on a huge scale. The surface of the moon has an area of 50 million square kilometers, and the radiations from the sun penetrate about

10 centimeters below the surface. Every month, therefore, the sun produces a permanent "sunburn" in 50 cubic kilometers of lunar soil. Furthermore, because of the continuous stirring of the soil, there are probably thousands of cubic kilometers of lunar material that have been "sunburned" during the past three billion years or so.

The lunar rocks and soil also contain traces of much stronger radiations whose energies are too great to have originated in our own sun. These high-energy radiations, called *galactic cosmic rays*, have been detected for many years in the earth's atmosphere and in freshly fallen meteorites; scientists believe that they must originate entirely outside our own solar system. The cosmic rays produce the same general effects as do the other radiations from the sun, but the higher-energy cosmic rays penetrate more deeply into the soil, producing detectable changes at depths of a meter or two below the surface.

SUNBURN ON THE MOON

"Sunburn" in lunar rocks and soil refers to permanent changes produced by the passage of high-energy atomic particles that continually bombard the moon. These particles originate in the sun and other stars, and they travel through space as streams of single atoms, most of which have lost electrons and become positively charged ions.

Three major kinds of particles can be distinguished on the basis of their source and their energy. The solar wind consists of low-energy particles emitted continuously from the outer atmosphere of the sun. Most of the particles are nuclei of hydrogen and helium atoms, but heavier atoms are also present. Solar wind particles have energies of a few thousand *electron volts*, or about that of the electrons in a TV tube.* Solar flares, periodic eruptions on the sun, produce occasional streams of particles with energies of hundreds of millions of electron volts. The most energetic particles, found in galactic cosmic rays, have energies of billions of electron volts, which approximates the energy of particles produced in our larger "atom-smashers."

The "sunburn" effects are produced by the collision of these high-energy atomic particles with the atoms that form the crystals and glass fragments in the lunar soil. Two kinds of effects are produced: nuclear reactions and particle tracks. Nuclear reactions occur when a high-energy particle of hydrogen or helium strikes the lunar atoms and changes them into atoms of different elements. Particle tracks are produced by heavier atoms, often as heavy as iron; the tracks consist of a permanent microscopic trail which marks the passage of the high-energy atom through the crystal.

Different "sunburn" effects are produced at different depths within the soil, depending on the energy of the particle involved. The higher the energy, the deeper the penetration into the soil (see Table 1). The low-energy solar wind particles do not travel more than one-thousandth of a millimeter into solid material, and they are absorbed on the outer surfaces of tiny grains. The higher-energy solar flare particles penetrate much farther, about 10 centimeters into the soil, while the highest-energy cosmic rays can produce detectable nuclear reactions a few meters below the surface.

THE SOLAR WIND

The solar wind is a stream of atoms emitted continuously out of the upper part of the sun's atmosphere. In this part of

* One *electron volt* is the energy given to a single electron by accelerating it through a potential difference of 1 volt. It is a very small amount of energy, less than one billion-billionth of a calorie.

Table 1 / "Sunburn" Effects in Lunar Rocks and Soil

SOURCE AND ENERGY OF PARTICLES	NATURE OF PARTICLES	EFFECT PRODUCED BY PARTICLES	MAXIMUM DEPTH OF EFFECT
Solar wind low energy (about 1,000 ev*)	Light atoms (hydrogen and helium) dominant, rarer heavier atoms (carbon, nitrogen, oxygen, etc.)	Atoms trapped in amorphous surface layer of lunar dust grains; chemical reactions	Less than 0.001 mm*
		Very small particle tracks	Less than 0.001 mm*
Solar flares *high energy* (1–100 million ev)	Light atoms (hydrogen and helium) dominant, rarer heavy atoms (e.g., calcium, iron)	Nuclear reactions**	About 6 cm*
		Particle tracks**	About 3 mm
Galactic cosmic rays very high energy (1–10 billion ev)	Light atoms (hydrogen and helium)	Nuclear reactions**	1–2 meters
	Heavy atoms (e.g., calcium, iron)	Particle tracks**	About 10 cm

* ev = electron volts; mm = millimeter (about 1/25 inch); cm = centimeter (10 mm).

** indicates effects most commonly used for measuring exposure ages in lunar samples.

the sun, temperatures are so high that the atoms move fast enough (several hundred kilometers a second) to overcome the gravitational field of the sun. Because the atoms are charged, they are accelerated into space by the sun's magnetic field, and eventually they sweep over the earth, the moon, and the other planets. The solar wind is a relatively new discovery. Its existence was predicted in the late 1950s by scientists studying the sun, and it was detected by unmanned satellites launched several years later.

The streams of solar wind atoms move through space at about 500 kilometers per second. This is many times the speed of a spacecraft going to the moon, but it is a slow crawl compared to the speed of light—300,000 kilometers per second. Light travels the 150,000,000 kilometers (93,-000,000 miles) from the sun to the earth in about 8½ minutes, while the solar wind takes about 2 days to make the trip.

Compared to the near-vacuum of space, the solar wind contains a great deal of material, usually 2 to 5 atoms in each cubic centimeter. Near the orbit of the earth, several hundred million particles of the solar wind strike each square centimeter of surface every second. None of the material reaches the earth's surface; most of the solar wind is deflected by the earth's strong magnetic field, and a small portion is trapped in the upper atmosphere. On the moon, which has neither atmosphere nor a significant magnetic field, all of the solar wind strikes the surface and is trapped in the soil. In the 3½ billion years since the lunar maria were formed, the moon has accumulated several *trillion* tons of matter from the sun, carried there by the solar wind. (This is only about one ten-millionth of the weight of the moon, so the moon is hardly putting on weight by consuming the solar wind.)

Because the sun is composed almost entirely of hydrogen and helium, these are the atoms most common in the solar wind. The Apollo Solar Wind Composition Experiment (see p. 92) also detected small amounts of heavier atoms such as neon, argon, krypton, and xenon; and analyses of the lunar soil have further identified carbon, oxygen, and nitrogen as additional trapped components of the solar wind. The Apollo experiments showed that the ratio of helium to hydrogen in the solar wind was only about 0.04, whereas in the upper atmosphere of the sun ratios of about 0.1 have been measured. This suggests that some process in the upper atmosphere of the sun serves to hold back the helium atoms and eject the hydrogen atoms preferentially into space.

The low-energy solar wind atoms are trapped in small

lunar soil particles and do not penetrate any deeper than about one-thousandth of a millimeter. The atoms of gaseous elements, such as helium and the other noble gases, can be removed by heating the soil, and scientists who made the experiments on the first returned soil samples were astonished to discover how much gas the lunar soil contained. When melted in the laboratory, the samples often released amounts of gas equal to their own volume. Careful microscopic studies of the soil particles showed that much of the gas occurs as very tiny high-pressure bubbles in crystals. Studies of these solar wind gases are important because they are a record of the sun's activity over millions, and possibly hundreds of millions of years.

In the outermost parts of small crystals and glass fragments in the lunar soil, the solar wind atoms also produce tiny particle tracks less than one-thousandth of a millimeter long. These tracks form so rapidly that a single grain may contain billions of tracks in each square millimeter of area. Sometimes the particle tracks are so numerous that the outer surface of the grain is transformed into an amorphous material whose properties are quite different from those of the original grain. The exact nature of this amorphous layer is the subject of much excited study, for its importance may extend beyond the moon. The outer layer may help small particles in space stick together to form larger ones and thus begin the whole process of planetary accretion. The amorphous surfaces may also provide the catalyst needed to form more complex chemical molecules out of the atoms of the solar wind, thus taking the first uncertain steps along the path that may lead to life.

SOLAR FLARES AND COSMIC RAYS

The solar wind is a steady, low-energy phenomenon whose effects on lunar rocks are mild and confined to a thin surface layer. By contrast, solar flares are sudden, violent eruptions on the face of the sun. Solar flares spray the lunar surface with bursts of high-energy particles and produce permanent and detectable changes to depths of several centimeters in the lunar soil. A typical solar flare is brief, usually lasting from a few minutes to a few hours, but it may cover an area on the sun thousands of kilometers across and produce temperatures as high as 40,000° C., many times higher than the sun's normal surface temperature of about 6,000° C.

The atomic particles that are ejected from the sun by solar flares are the same as those in the solar wind, mainly hydrogen and helium with traces of heavier elements. But solar flare particles emerge from the sun with much higher energies, and they travel at speeds much closer to that of light—often as fast as 150,000 kilometers per second. Even these energetic particles cannot penetrate the protective shield of the earth's atmosphere, but they are strong enough to disturb the atmosphere's electrical properties and to disrupt radio communications. In space, beyond the protection of the atmosphere, solar flare particles can penetrate spacecraft and create a potential hazard for astronauts, just as intense X-rays or radioactivity are dangerous to life on the earth's surface.

On the lunar surface, solar wind particles can penetrate several centimeters into the lunar soil. Particle tracks produced by solar flare atoms can be detected in the outer few millimeters of lunar rocks, and the abundance of such tracks may reach several hundred million in a millimeter-size grain. Nuclear reactions can be produced by solar flare particles at depths of several centimeters in the lunar soil.

The still higher-energy galactic cosmic rays are similar to solar flare particles and also consist of energetic atoms. However, the cosmic ray particles travel much faster, nearly at the speed of light, and they contain several times as much energy as solar flare particles. Their energies, measured in billions of electron volts, are too large for them to have been produced in the sun, so they must come from outside of the solar system, possibly from other stars or from the interstellar regions of our own Milky Way Galaxy. The cosmic rays must have been traveling for thousands, or possibly tens of thousands, of years before striking the moon. Very little is known about these strange and powerful particles that are strong enough to produce nuclear reactions a few meters deep in the lunar soil, and the records they leave are an exciting source of information about the history of the universe outside of our own solar system.

Earthbound scientists have been studying cosmic rays with sensitive detectors and instrumented satellites for many years, but one of the major surprises of the Apollo Program was the unexpected discovery that human beings in space could detect cosmic rays without any instruments at all. The Apollo 11 astronauts reported that, during the mission, they occasionally "saw" bright streaks of light while their eyes were closed. Careful studies on later missions established that the streaks were caused by the passage of high-energy atomic particles through the astronauts' eyeballs. The human eye turns out to be a fairly sensitive detecting instrument for cosmic ray particles. With their eyes closed, many astronauts could "see" one streak about every three minutes while they were in space between the earth and the moon

A more conventional measurement of the exposure of the astronauts to cosmic rays was made by studying their plastic space helmets. A cosmic ray particle passing through a helmet leaves a tiny invisible trail of damaged plastic. These tracks are etched out by treating the helmet with certain chemicals, and appear as tiny holes in the plastic (photo 29). Counting the holes gives the number of cosmic rays that passed through the helmet during the mission.

These examinations indicated that astronauts were exposed to increasing numbers of cosmic rays during the later missions. The helmets from earlier missions, such as Apollo 8 and 11, contained about one track in each two square centimeters of surface, or about 0.5 track per square centimeter. Helmets from the later missions showed about 3 tracks per square centimeter, even after corrections were made for the fact that the later missions were longer.

The apparent increase in cosmic rays during the later Apollo missions may have been due to several factors. It is possible that on the earlier missions the helmets may have been stored in a more protected part of the spacecraft where fewer cosmic rays could reach them. Another explanation is that the sun's activity, which follows a regular 11-year cycle, was decreasing slightly during the later missions. As the intensity of the sun's magnetic field decreased, more cosmic rays could penetrate from outside the solar system into the region occupied by the earth and moon.

These studies of cosmic rays were important for medical as well as scientific reasons, because the high-energy atomic particles of cosmic rays can penetrate and destroy the living cells of animals and humans. Before the Apollo missions, our data indicated that cosmic rays would not be a hazard on the relatively short trips to the moon. But with men actually in space, more accurate measurements of cosmic ray exposure could be made, and scientists could estimate better the possible dangers for the Apollo astronauts and for future space travelers.

The data from the helmets and the eyeball flashes were reassuring—for the moment. Cosmic rays are no danger to astronauts during the few weeks of a lunar mission. But during a longer stay in space, such as a 2-year trip to Mars, the cosmic ray exposure could be a significant hazard, and more protection for the spaceship and the astronauts would be needed.

HOW TO READ A SUNBURN

"Sunburned" lunar rocks are much like sunburned people. In both cases, exposure to the sun produces distinctive effects, regardless of whether the specimen is lying on the lunar surface or on a terrestrial beach. From the permanent effects of lunar "sunburn" in rocks and soil, geologists can determine when a rock was exposed, how long it was exposed, what position it was in, and how much cover it had. From this information, they can reconstruct both the record of movement of the lunar soil and the past history of the sun.

The type of sunburn acquired by a lunar rock depends on how deeply it is buried in the lunar soil (see Table 1). Below about 2 meters virtually no outside radiations penetrate, so rocks and soil below this level acquire no measurable sunburn. The upper 2 meters of the soil can be divided into three layers in which different kinds of sunburn develop: a lower zone from 2 meters to about 10 centimeters deep, a "sub-decimeter" zone from 10 centimeters to the surface, and the surface itself.

Lunar rocks and soil located less than 2 meters from the surface are penetrated by high-energy cosmic ray particles that strike the atoms of the lunar material and produce nuclear reactions. The reactions create new atoms, and the longer the sample is exposed to cosmic rays, the more new atoms are created. Scientists can determine what is called an "exposure age" by measuring the amounts of any new atoms that have formed in the specimen. Cosmic rays produce about 20 different kinds of atoms in lunar rocks, all of which are isotopes of known elements. Some of the more common isotopes used to measure exposure ages are: neon-21, aluminum-26, argon-39, nickel-59, and krypton-81. Using different elements, several independent exposure age measurements can be made on the same sample, thus providing a check on the accuracy of the exposure age obtained.

More shallow samples acquire different "sunburn" effects. A sample less than about 10 centimeters from the surface will contain particle tracks caused by the passage of the high-energy cosmic rays. A particle track is a trail of atomic-scale damage produced as the particle smashes its way through a mineral fragment or a piece of glass. Just like the tracks in the Apollo helmets, the tracks in lunar rocks can be etched into visible holes by using suitable chemicals. The tracks are only a few hundredths of a millimeter long, but they can be easily observed and counted under the microscope.

Examination of the tiny grains in lunar soil has shown that they are literally riddled with cosmic ray particle tracks. A grain only a millimeter square may contain millions of separate tracks. The longer a grain or rock fragment remains at a depth of less than 10 centimeters from the surface, the more particle tracks it will accumulate, and from the number of tracks, scientists can calculate a second kind of exposure age that measures the time that the sample has spent close to the surface of the moon.

Samples in this "sub-decimeter" (less than 10 centimeters) zone also show nuclear reactions produced by the lower-energy atomic particles from solar flares. Within the upper centimeter or so of this layer, the solar flare atoms also produce particle tracks.

Fortunately, it is possible to distinguish both the nuclear reactions and particle tracks produced by solar flares from those that are produced by higher-energy cosmic rays. One method for separating the effects is based on the fact that the solar flare particles do not penetrate as far into solid matter as do the more energetic cosmic rays. On instructions from the scientists, the astronauts on several missions collected large rocks about the size of a grapefruit. Solar flare particles penetrate only part way into such large rocks, while the more powerful cosmic rays pass right through them. By comparing the numbers of tracks in the center of a rock with the numbers of tracks observed near the surface, it is possible to sort out the effects of solar flares from those of cosmic rays.

Finally, a rock on the surface of the moon acquires still a third kind of sunburn consisting of many small tracks produced by low-energy solar wind particles which penetrate only a fraction of a millimeter into rocks and soil. Both microcraters caused by bombardment by cosmic dust particles and solar wind particle tracks appear only on the surface of the moon and only on the upper sides of exposed rocks.

The sunburn in a single lunar sample is the end result of several complicated processes. Sunburn depends on how fast the lunar soil is stirred and mixed, on how deeply a specimen is buried, and on how long the specimen remains buried at different depths. The stirring of the lunar soil depends on the rate at which solid particles strike the moon, but this same rate of bombardment also determines how long a sunburned rock, exposed at the surface, will survive before being eroded by small particles or broken apart by larger ones. As if this complexity were not enough, lunar sunburn will also reflect any changes in cosmic rays or the sun's intensity during the billions of years in the past.

In order to unravel these relationships, the study of sunburned lunar rocks goes on in stages, beginning with measurements and ending with calculations and theories. Scientists first determine exposure ages on many rocks and soil samples. From the exposure ages, the movement of the lunar soil can be hypothesized. Then, the calculated movements of the lunar soil provide estimates of the bombardment rate. Finally, as a great amount of data continues to accumulate, the past history of the sun and galactic cosmic rays is beginning to appear.

EXPOSURE AGES: THE TRAVELS OF A LUNAR ROCK

Even the first exposure ages measured on returned Apollo 11 samples confirmed the old idea that things move very slowly on the moon. The rocks collected on the lunar surface have been within at least 2 meters of the surface for periods ranging from 10 million to 500 million years. The rocks collected on later missions had similar exposure ages, so it is now firmly established that rock fragments can remain intact at the surface of the moon for tens or hundreds of millions of years. However, the rocks themselves are many times older, with measured formation ages of between 3.3 and 4.2 billion years. The rock fragments, despite their high exposure ages, have nevertheless spent only a small fraction of their lifetime as broken pieces near the lunar surface.

The long survival of rocks at the moon's surface confirms what was indicated by the study of the salvaged Surveyor 3 parts—namely, that erosion on the moon is a slow process. It takes several million years for the bombardment of small cosmic particles to wear off a layer only a millimeter thick from the surface of a freshly exposed lunar rock. Even so, the smoothed and rounded upper surfaces of many exposed rocks are eloquent testimony to the millions of years they have spent sitting undisturbed on the moon.

Exposure ages measured on samples of lunar soil vary greatly because the soil is a mixture of old and new materials that have been exposed for widely different lengths of time. The average exposure "age" of lunar soil samples, about 400 million years, is similar to the exposure ages measured on many individual rock fragments. However, some fine particles in the lunar soil have been exposed for longer periods of time. The exposure ages determined for some tiny mineral grains by measuring particle tracks are as high as 1,700 million years. This indicates that large rocks do not remain intact perpetually on the lunar surface. After a billion years or so, they are broken down into smaller grains that subsequently remain in the upper part of the soil layer.

Exposure ages of rocks can also be used to date the formation of craters on the lunar surface. Even a relatively small crater 50 to 100 meters in diameter is deep enough to penetrate through the lunar soil layer, and the formation of such a crater brings up large quantities of fresh bedrock that has never been exposed to the sun. The exposure ages of these excavated fragments of bedrock thus indicate when the crater formed.

Many of the Apollo 11 samples have exposure ages close to 100 million years. This, for example, may be the age of West Crater near the landing site. On later missions, scientists were able to link samples to specific craters with greater certainty. Many Apollo 14 samples have exposure ages of 25 to 30 million years, which is almost certainly the age of Cone Crater. The two most prominent craters in the Descartes landing site (Apollo 16) are also well dated. The older, more subdued North Ray Crater is about 50 million years old, while the younger, fresher South Ray Crater formed only 2 million years ago. Shorty Crater in the Littrow Valley (Apollo 17) formed about 30 million years ago, exposing the Orange Soil. The exposure ages of single rocks from the Apollo 15 and 17 missions indicate that even younger impacts occurred less than a million years ago, although the craters themselves have not yet been identified.

Despite the slowness of movement on the lunar surface, some lunar rocks have managed to travel far from their original locations, and the sunburn effects in the samples have kept an accurate diary of their wanderings. The sunburn record traces the stages in a rock's journey from buried bedrock up through the lunar soil to the surface, and can even follow the rock's movements on the surface as it was flipped and spun by small meteorite impacts every few million years or so.

The life history of sample 10017, one of the larger rocks returned by Apollo 11 from Mare Tranquillitatis, spans the history of the earth and the human race as no earthly record does. Rock 10017 was formed from molten lava about 3.6 billion years ago and remained buried deep below the lunar soil for more than 3 billion years. About 500 million years ago, when complex animals first began to appear in the earth's oceans, a meteorite impact shattered the bedrock and threw the rock fragment out into the lunar soil. For nearly half a billion years, rock 10017 lay buried between 10 centimeters and 2 meters deep in the lunar soil, while the whole pattern of life flourished and changed on earth. Six million years ago, long after the dinosaurs had come and gone, another impact threw rock 10017 onto the surface of the moon. It was flipped upside down about 3 million years later, just about the time that a small group of ape-like animals in Africa developed tools and began the changes that would lead to man. Three million years after that, one of their remote descendants, dressed in a spacesuit, landed on the moon, picked up rock 10017, and brought it back to earth.

THE COSMIC HAILSTORM: CLOUDBURST OR DRIZZLE?

The rock fragments exposed on the lunar surface have been pivotal in determining how often the moon is being struck by large and small cosmic particles from outer space. Before the Apollo landings, our estimates of the cosmic bombardment rate came from telescopic crater counts of the moon and from instruments on artificial satellites. Both of these methods were uncertain; the ages of the lunar surfaces were not well known, and the satellite instruments often gave false signals, producing calculated bombardment rates that were eventually shown to be as much as a factor of 100 too high.

The exposed surface of a lunar rock is a natural particle counter. The tiny, fast-moving particles of cosmic dust that strike the rock produce tiny, glass-lined microcraters. If the exposure age of the rock is known, we need only count the craters on the exposed surfaces to calculate the rate at which the rock has been bombarded.

The Apollo missions secured evidence that the bombardment of the moon by cosmic dust goes on today, because the Apollo spacecraft themselves were repeatedly (but harmlessly) struck by tiny particles. A thorough examination of the windows of the Apollo spacecraft after they returned to earth disclosed about 10 tiny craters, the largest of which was only half a millimeter in diameter. The cratered lunar rocks provide a far longer record, extending back over thousands and millions of years.

So far as we know, the impact rate of small dust particles onto the moon has been constant throughout the past few million years. However, some recent studies have raised questions about this assumption. Scientists who made a complete study of the microcraters on a large Apollo 16 sample (60015) (see photo 26) were able to date the formation ages of single microcraters by studying the solar flare particle tracks preserved in the glass linings of each crater. The rock had been exposed on the lunar surface for about 80,000 years, but the data indicate that more microcraters had formed during the last 10,000 years than in the earlier stages of the rock's exposure.

Such a recent increase in microcrater formation would imply a sudden influx of dust-size particles into the earth–

moon region about 10,000 years ago. One possible explanation is the arrival of a new comet which began to shed dust as it passed close to the sun. Comet Encke, first detected in 1795, which circles the sun every 3.3 years, is a good possibility for the new source of dust because it passes by the orbit of the earth so frequently.

But there is still not enough data to determine whether this recent increase is a real effect; more data from new samples will be needed. At present, however, there is an exciting possibility that lunar rocks could be used as a "comet detector" that records the passage of these strange and poorly understood bodies.

THE LUNAR ATMOSPHERE: WHAT KIND OF NOTHING?

An *atmosphere* is a gaseous envelope which exists above the solid surface of a planet, held close to it by the force of gravity. The nature and chemical composition of the atmosphere provide important data about the origin and evolution of the planet itself. Atmospheres are not difficult to study. Even the atmospheres of distant planets like Mars and Jupiter can be analyzed with earth-based instruments, while instrumented spacecraft can provide even more detailed information about planetary atmospheres from distances of millions of miles.

The gases that form a planet's atmosphere may come from many sources: from the solar nebula during the formation of the planet, from the interior of the planet, from the activities of life, or from the solar wind. The hydrogen and helium that form most of the atmosphere of the giant planet Jupiter are probably original, held for billions of years by the strong gravity of the planet. By contrast, it appears that virtually all of the earth's atmosphere was formed by gases that came out of the earth's interior during long episodes of volcanic activity. Such an origin seems likely for the nitrogen that makes up four-fifths of our atmosphere, the water in our oceans, and the carbon dioxide and water that are present in tiny amounts in our air. Oxygen, which makes up one-fifth of our atmosphere and supports the life on our planet, probably was produced from original carbon dioxide by the activity of plant life during the course of billions of years. Jupiter's atmosphere may contain the original material from the solar nebula, but the composition of earth's atmosphere testifies to the chemical modifications made by life.

The moon is completely different from the larger planets. For all practical purposes, the lunar atmosphere is nonexistent. However the space above the surface of the moon is not entirely occupied by nothing. Near the lunar surface there is a tenuous swarm of gas atoms that can be detected only with sensitive instruments. The pressure of this lunar "atmosphere" is not even one ten-trillionth that of the earth's, but even so, there are still several million atoms in each cubic centimeter.*

The abundance and nature of these atoms can tell us much about what is happening both on and within the moon. Light atoms, such as hydrogen and oxygen, cannot be retained long by the moon's weak gravity; if they are

* By comparison, a cubic centimeter of air at the surface of the earth contains about 25 million trillion atoms. In fact, the weight of all the matter in the lunar atmosphere is less than about 50 tons, and the whole lunar atmosphere, if frozen, could be packed inside a single moving van. By contrast, each Apollo Lunar Module contained 12 tons of propellants for its two rocket motors, and much of this propellant was burned near the lunar surface during landings and takeoffs. Each Apollo mission thus increased the lunar atmosphere significantly, perhaps as much as 10 to 20 percent. The gases added—mostly water vapor and nitrogen oxides—could be detected during takeoffs by the instruments placed on the lunar surface.

These man-made additions to the lunar atmosphere are probably only temporary. The larger molecules will be decomposed by the intense sunlight and ultraviolet radiation near the lunar surface, and the lighter atoms and water vapor will soon be lost to space.

present in the lunar atmosphere, they must somehow be continually supplied to the lunar surface, either from the solar wind or from some source inside the moon. However, heavier gases, such as argon and krypton, could have been held by the moon since it formed. The presence and composition of even tiny amounts of lunar gases could tell us much about the moon, so the Apollo missions put sensitive detectors onto the lunar surface and carried others into lunar orbit to find out just what kind of gases were there.

The detectors found that the tenuous lunar atmosphere consists chiefly of the elements hydrogen, helium, neon, and argon. The instruments also detected several gases that were clearly man-made. Nitrogen oxides were produced by the rocket engines during landing and takeoff. The surface instruments, which were sensitive enough to detect a few hundred molecules of gas, easily recorded the tiny, harmless leaks of water vapor and oxygen from the spacesuits of the astronauts as they passed by. Finally, bursts of water vapor and other gases were released by the abandoned descent stage of the Lunar Module for weeks and months after the astronauts had returned to earth. Fortunately, these man-made contributions freeze out of the atmosphere during the chill lunar night, so that nighttime measurements were able to determine the true nature of the lunar atmosphere.

The tiny amounts of lunar gases that were detected apparently come from several sources. Some of the gases, especially argon, are diffusing slowly out of the lunar interior and may provide clues to processes going on within the moon. The solar wind brings a steady stream of atoms of hydrogen, helium, carbon, nitrogen, and oxygen to the lunar surface, and if the atoms react chemically to form gaseous molecules, such reactions supply another material to the atmosphere. Finally, meteorite impacts can also build up the lunar atmosphere, either by vaporizing lunar rock or by releasing any extraterrestrial gases contained in the meteorite itself.

Furthermore, a small part of the lunar atmosphere is definitely produced inside the moon. Much of the element argon in the lunar atmosphere is present as the isotope argon-40 and has probably been produced by the decay of radioactive potassium-40 in the rocks inside the moon. This argon gas has slowly made its way from the inside of the moon to the surface, accompanied by small amounts of radon gas produced by the decay of other radioactive elements (see pp. 114–15).

The amount of gas diffusing out of the lunar interior is too small to suggest that any intense volcanic activity is now going on within the moon. One possible indicator of volcanic action is the reported emission of water vapor detected by the instruments on the lunar surface. The possibility that the water is being emitted from the abandoned descent stage of the Lunar Module cannot be ruled out entirely. Even if these water-producing events are natural, they are uncommon, and the amount of gas involved is too small to indicate major volcanic activity.

There is stronger evidence that gases were emitted from the lunar interior in the ancient past. Some kind of gas was clearly associated with the lavas that flooded the maria about 3½ billion years ago, for we commonly find gas bubbles preserved in the solid rock (photo 30). It is unlikely that this gas contained water; no water has ever been found in any of these bubble-rich lunar rocks. Possible candidates for the gas are hydrogen, carbon monoxide, hydrochloric acid, hydrogen sulfide, and similar gases also common in terrestrial volcanoes.

Some of the lunar volcanic rocks collected on later Apollo missions contain traces of volatile materials which could have been present when they were erupted. Specimen 66095, the so-called "Rusty Rock" collected by Apollo 16 astronauts, contains the only water so far detected in a lunar rock. The water is tied up in small amounts of red-orange iron hydroxides which occur as streaks and coatings in the specimen, and which closely resemble terrestrial rust. The lunar origin of this "rust" and the water it contains are not yet accepted by everyone, but it is significant that the specimen is also rich in other volatile elements like zinc and chlorine, which could have been present with the water in a lunar volcanic gas. The Orange Soil collected by Apollo 17 astronauts is also rich in zinc, chlorine, and other volatile elements, suggesting that the Orange Soil was erupted as a gas-rich "fire fountain," or volcanic fumarole, on the moon.

Although gases from the lunar interior may have been important in the moon's distant past, nearly all of the moon's present-day atmosphere is produced by the solar wind, which apparently provides all the hydrogen, helium, and neon detected on the lunar surface and much of the argon as well.

When these solar wind atoms are trapped in the lunar soil, chemical reactions between the trapped atoms and the minerals in the soil produce small amounts of other gases, which have been found by analyzing the lunar soil. These reactions are especially active in the uppermost layer of the lunar soil, which is heated above the boiling point of water during the lunar "day." Atoms of carbon and hydrogen combine to form methane (CH_4). Oxygen and carbon atoms react to produce carbon monoxide (CO). Hydrogen atoms from the solar wind react with sulfur-bearing minerals in the soil to produce hydrogen sulfide (H_2S).

Studies indicate that once they are formed, these gas molecules diffuse slowly out of the lunar soil to the surface, where they are either destroyed or escape into space. The amounts of these gases are small, but their loss from the soil removes detectable amounts of carbon, oxygen, and sulfur, producing slight but detectable changes in the chemical composition of the soil.

Additions to the lunar atmosphere by meteorite impacts are less certain and probably temporary. A small meteorite impact that melts some lunar soil will release a sizable amount of trapped solar wind gases, but by the time these gases are dispersed into the lunar atmosphere, the effect will be hardly detectable, and much of the released gas will eventually be lost from the moon. A larger meteorite impact that vaporizes several cubic kilometers of rock will provide a brief, hot atmosphere of silicon vapor and other volatile elements, but this material quickly condenses on the lunar surface and is incorporated into the lunar soil.

It has been suggested that the impact of a very large, volatile-rich meteorite or comet would release a great deal of gas into the lunar atmosphere; such an extraterrestrial source has been proposed to explain the water, chlorine, and other volatile elements in "Rusty Rock" and in the Orange Soil. At present, no volatiles of definite meteorite or cometary origin have been identified in lunar samples, and it seems unlikely that such large and infrequent impacts have contributed more to the lunar atmosphere than the continuous influx of the solar wind.

Because of the high vacuum and the high daytime temperatures on the lunar surface, the daytime lunar atmosphere may contain vapors of many elements that are normally solid on earth. Samples of lunar soil collected from permanently cold and shadowy places under large boulders seem to be enriched in elements like lead, mercury, bromine, antimony, and other elements that have low boiling points. These data suggest that such volatile elements are distilled out of the lunar soil by the hot sun, and then move into the cold shadows where they condense again during the lunar night. This migration of elements produces a particular problem in the case of lead, which can be produced by radioactive decay and then migrate far away from its parent elements, thus making age determinations on near-surface lunar samples very difficult to interpret. Another interesting aspect of this movement of metal atoms across the moon is the possibility that areas near the lunar poles, which have been in permanent shadow, may contain sizable deposits of metals or unique minerals formed by the condensation of volatile elements over billions of years.

FROM THE MOON TO THE STARS: THE TRACKS OF THE SUN

Before the Apollo landings, an important source of direct information about the sun was freshly fallen meteorites, sometimes referred to as "the poor man's space probe," because they had been exposed to solar radiation while in space. Unfortunately, the meteorite record is incomplete, because as meteorites make their fiery passage through the earth's atmosphere, their outermost surface is burned off; and with this outer layer goes the record produced by very low-energy solar particles that do not penetrate deeply into solid matter.

Therefore, the Apollo landings on the moon were also a great leap forward for the study of the sun. Not only were we able to collect samples of the solar wind to analyze the matter coming out of the sun today, but more importantly, because the slow erosion of solid rocks on the moon removes very little surface material, the returned lunar rocks provided a far more complete record of the past history of the sun than our previous meteorite source.

Almost immediately, scientists learned to their astonishment that the material ejected from the sun by intense solar flare eruptions is not the same as the overall composition of the sun itself. The ratio of hydrogen to helium atoms in solar flare particles turned out to be about 20 to 1, although the ratio of such atoms in the sun itself is about 10 to 1. The poorly understood mechanisms of solar flares seem to favor the ejection of the lighter hydrogen atoms from the sun.

However, examinations of the solar flare particle tracks etched into the Surveyor 3 TV camera lens during its 2½-year exposure on the moon yielded somewhat contradictory information. These studies showed that the iron-to-hydrogen ratio in solar flares is 10 times higher than in the sun itself, indicating that the solar flares in this case preferred to eject the heavy iron atoms instead of the light hydrogen. Clearly, solar flare formation is more complex than we had thought; perhaps different solar flares even have different chemical properties.

The lunar samples also allowed us to answer questions about the past behavior of the sun. Was the sun more or less active in ancient times? Have solar flare eruptions always been present, or are they a fairly recent phenomenon that prefigures some unexpected change in the overall behavior of the sun? These are not idle questions; a change of only 3 percent either way in the amount of energy put out by the sun would make the earth uninhabitable. Smaller changes in the sun could cause climatic changes on the earth severe enough to threaten not only our present civilization but our survival as a species. The present and future behavior of the sun is an intriguing scientific question but it is also one in which everybody has a personal stake.

So far, the conclusions are comforting. The record in the lunar samples indicates that the behavior of the sun has been almost surprisingly constant over periods of time ranging from thousands to billions of years. For example, the composition of the solar wind atoms trapped in lunar soils shows no measurable changes over about a billion years.

Similarly, solar flares seem to have occurred 100,000 years ago just as they do now. Modern solar flares have left traces in many lunar samples, and the effects of these modern flares provide an important standard for reading the more ancient records. For instance, a massive solar flare on January 24, 1971, produced both nuclear reactions and particle tracks in the upper part of a boulder collected near Hadley

Rille by the Apollo 15 astronauts seven months later. The effects were located on the top side of the rock, showing that the boulder had not moved between the time of the flare and the time it was collected.

Several lines of evidence point to a constancy of solar flare activity in the past. The amount of the isotope argon-39 in lunar soils suggests a uniform occurrence of solar flares for at least the past 1,000 years. Other analyses indicate that the number and energy of helium atoms emitted in flares has remained constant for 80,000 to 100,000 years and that the ratio of hydrogen to helium has been constant at about 20 to 1 for the same length of time. Finally, studies of particle tracks indicate that the abundance of iron atoms in solar flares, compared to the number of iron atoms in galactic cosmic rays, has been constant for at least 270,000 years.

If the sun seems to have behaved constantly for the last 300,000 years or so, we may have some confidence that it will continue to be the same for at least a few centuries more. But it seems premature to assume that the sun has always been the way it is now. Some specimens of highland rocks and minerals contain unusually large numbers of particle tracks, and one possible explanation is that the sun was far more active 4 billion years ago than it is now. This question, like so many others, is still unanswered, but it would be exciting if the same lunar rocks which also reassure us about the sun's present stability could also provide a record of the sun's wilder early days.

The apparent constancy of the sun seems matched by a similar constancy in the rest of the universe, for the record of the galactic cosmic rays also seems to have been the same for nearly the last billion years. Studies of meteorites and lunar samples together have shown that the ratio of iron to calcium atoms in cosmic rays has remained the same over this period; other studies indicate that the abundance of high-energy hydrogen atoms in galactic cosmic rays has not changed for at least the last billion years.

In one way, this constancy of cosmic rays is somewhat surprising. The Milky Way Galaxy, in which our sun is located, has made about five complete revolutions around its center during the last billion years, while our sun and the planets speed through the galaxy at about 20 kilometers per second. In the last billion years, the sun should have passed through many different regions of the galaxy, and the apparent constancy of cosmic rays during all this time is an important discovery. It may mean that the cosmic rays are produced by a single mechanism everywhere in the galaxy. Or it may mean that the sun has spent the last billion years in a part of the galaxy where cosmic ray intensity is very uniform. One of the biggest challenges for the future is to read in the lunar samples the full record of these few atoms that were born in distant stars and traveled across space for thousands of years before ending their journey in the surface of the moon.

A RECIPE FOR LIFE

In many respects, the worst place to search for the origin of life is on a planet like the earth where life has been thriving for billions of years. The original chemicals necessary to create life have all been used up or transformed into more complicated molecules out of which modern animals and plants are made. The first simple living things have been superseded by far more complicated descendants. Even the original atmosphere has been modified by life, for the oxygen that we breathe is the by-product of the activities of plants for hundreds of millions of years.

The Apollo missions, while demonstrating beyond any doubt that the moon is (and always was) lifeless, also gave us our first opportunity to study a planet whose original nature had never been modified by life. It well may be that the chemical reactions we detected between atoms of the solar wind and particles of lunar dust duplicate the first simple chemical steps that had to occur before life could develop, yet the barren moon affords an example of an environment where life itself could not arise.

Even as a source of biologically useful chemicals, the lunar soil is disappointing. Carbon, the essential element for life, is nearly absent. The lunar soil contains less than about 200 parts per million (0.02 percent) of carbon. The lunar rocks from which the soil has formed contain even less carbon, and it seems clear that nearly all the carbon in the soil does not come from the moon but has been added by the solar wind and meteorites. Very tiny amounts of organic* compounds such as methane (CH_4) and ethane (C_2H_6) totaling less than one part per million have been detected, but no "biological markers" suggestive of life (such as alkanes or amino acids) have been identified. Nor are there enough suitable chemicals in the lunar soil to nourish life. Even the plants grown in lunar soil back on earth had to have nutrient solutions added.†

The importance of the lunar soil in the study of the origin of life is that it provides a practical demonstration of how simple molecules can form when atoms of the solar wind react on the surfaces of lunar dust grains. Even before the Apollo Program, we knew that some reactions of this type must be going on in interstellar space, for our radio telescopes have detected the presence of a large number of

* The term "organic" refers to chemical compounds composed dominantly of carbon and hydrogen which make up living plants and animals. Because "organic" compounds can also be produced by chemical reactions in the absence of any life, as in the lunar soil, it is necessary to distinguish between "organic" compounds and "biogenic" compounds produced by the activities of living things. No biogenic compounds have been found in the lunar soil.

† Early reports that some terrestrial plants showed increased growth rates when grown in lunar soil have not been substantiated by later experiments. Almost everything that a plant needs for growth (water, organic compounds, potassium, etc.) has to be added to the soil to make the plants grow at all.

simple, but biologically important, molecules—such as water, formaldehyde (H_2CO), hydrogen cyanide (HCN), and ethyl alcohol (C_2H_5OH)—that have somehow formed in the cold spaces between the stars. Such molecules could not exist at the temperatures within stars, but they probably form everywhere in the universe where atoms of hydrogen, carbon, nitrogen, and oxygen can react at lower temperatures on the surfaces of tiny grains of interstellar dust. These compounds, formed without life, are the basic "building blocks" of more complex compounds like proteins and amino acids, without which life cannot develop.

Analyses of the lunar soil have detected many simple compounds that apparently formed by similar reactions between solar wind atoms and the tiny bits of crystals and glass in the lunar soil. Most of these compounds are inorganic: carbon monoxide (CO), carbon dioxide (CO_2), hydrogen sulfide (H_2S), and sulfur dioxide (SO_2). The small amounts of organic methane (CH_4) and ethane (C_2H_6) which have been detected probably formed by similar reactions.

There is another group of lunar compounds which are of special interest because they apparently can react with water to form amino acids, the basic material from which proteins are built. These "amino acid precursors," as they have been called, occur in such tiny amounts (a few parts per billion) that their nature and even their existence are still debated. But if they are present in the lunar soil, then a critical step toward the origin of life could be taken simply by bringing the lunar soil into contact with water.

Such molecules form by a simple mechanism which can occur practically anywhere in the universe. The surfaces of tiny dust particles in the lunar soil act as catalysts on which atoms of the solar wind combine to produce simple molecules, while the sun's heat supplies the energy needed for reaction. This process has also been duplicated in the laboratory by bombarding lunar soil samples with the atoms of simple elements. The amorphous coating produced on the dust particles by solar wind bombardment (see p. 193) apparently provides a highly reactive surface where the chemical reactions can take place rapidly. The same reactions produced in the laboratory and on the lunar surface can also occur on tiny dust grains in interstellar space.

It is a long way from producing simple molecules to producing life, but these simple molecules represent a critical step forward. If the reactions in the lunar soil *are* the first steps toward life, then the recipe for life is simple, and the ingredients are spread throughout the universe.

Evolution on the moon could not proceed past the simple substances, for if more complex organic molecules managed to form, they would be quickly destroyed by the harsh environment of the moon's surface. But the fact that such primitive but important chemical reactions can occur even in the destructive environment of the lunar surface suggests that they can also occur almost anywhere. In an environment like the earth, protected by an atmosphere and nourished with water, the primitive molecules were able to survive to form more complex ones, developing more and more complicated structures, until something emerged that we, its descendants, would call life. The gulf between the dead moon and the living earth may not be as wide as we once thought it was.

CHAPTER

THE LUNAR INTERIOR: MAPPING WITH ECHOES

To understand any planet, we have to learn about its interior, for the basic character of a planet is not determined on its surface. The major geological processes that modify the earth—volcanoes, earthquakes, the magnetic field, and the drifting of continents—are all driven by forces generated deep within it.

Even our deepest mines and oil wells penetrate no farther than a few kilometers into the outermost parts of the earth. Only ten kilometers beneath our feet pressures reach thousands of atmospheres* and temperatures climb to hundreds of degrees Centigrade—conditions that prohibit direct exploration by man.

Because we cannot explore the interior of the earth di-

* One *atmosphere* is the pressure exerted by the earth's atmosphere at sea level. It is equal to 14.7 pounds per square inch or approximately one kilogram per square centimeter. Pressure within the earth, which is produced by the weight of the overlying rock, increases about 300 atmospheres for each kilometer of depth.

rectly, we have had to depend on knowledge acquired at its surface. Geologists have sometimes been able to obtain samples of the interior by taking advantage of the geological processes which bring pieces of the interior to the surface. The formation of great mountain ranges like the Alps or the Appalachians, followed by millions of years of erosion, ultimately expose masses of rocks formed at depths of 20 to 30 kilometers. Molten lavas, which may form as deep as 150 to 200 kilometers, often break off pieces of the surrounding solid rock and carry them to the surface through erupting volcanoes.

These scattered masses and fragments of rock may tell us a great deal about the outermost shell of the earth, but they can tell us nothing about its deeper interior. Most of our knowledge about our planet's deep interior has come from sensitive instruments on the surface that record the tiny effects produced by major processes within the earth. For the last century, scientists have studied the variable distribution of rock masses within the earth by measuring small fluctuations in the force of gravity from place to place. The detection of seismic waves generated by deep earthquakes has helped identify the materials that make up the inside of the earth. Measurements of the heat flowing out of the earth have helped us understand the origin of volcanic activity. And we now know that the strong magnetic field at the earth's surface is the detectable product of poorly understood processes occurring in its very core.

The use of surface measurements to decipher the earth's interior has been a crucial part of the exploration of our native planet. Although new discoveries will continue to occur and new theories will arise to explain them, the general outlines of the earth's interior now seem fairly well established.

The earth is not a uniform, homogeneous body: its interior is made up of several different layers. The outermost layer, with which we are most familiar, is called the *crust*; it is composed of relatively light rocks like granite and basalt. (On the scale of the entire planet, the near-surface layers of sedimentary rocks like limestone and shale are insignificantly thin.) The thickness of the crust varies. Under the continents, where it is composed of granite and similar rocks, it may be 30 to 40 kilometers thick, while under the ocean basins, the crust is composed largely of basalt and is about 5 kilometers thick.

Beneath the crust there is a sharp transition to the heavier rocks the compose the earth's *mantle*.* The mantle is richer in iron and magnesium than the crust and is composed mainly of the minerals pyroxene and olivine. Nearly 3,000 kilometers thick, the mantle extends from the bottom of the crust to the outermost part of the earth's core. At this point, about 2,900 kilometers below the surface, there is another fundamental boundary; the silicate rocks of the mantle give way to the denser metallic material that makes up the earth's core.

The innermost part of the earth is a spherical metallic core about 6,000 kilometers in diameter—nearly half the diameter of the earth itself. Data from earthquakes and from chemical models of the earth suggest that the core is mostly metallic iron, probably alloyed with 10 to 15 percent of a lighter element like silicon or sulfur. The outer part of the core is in a molten state, and it is likely that motions in this outer core produce the earth's magnetic field. The inner core is solid as a result of the tremendous pressures (nearly a million atmospheres) present at these depths.

During the last few years, geologists have developed a

* The boundary between the earth's crust and mantle is named the *Mohorovičić discontinuity* (generally called the "Moho" for short) after the Yugoslav scientist who discovered it in 1909 by studying the behavior of earthquake waves. The Moho is a fundamental boundary within the earth; the rocks above and below it are significantly different in chemistry, mineral composition, and physical properties. The Moho is 50 to 60 kilometers deep under the continents, but only about 5 kilometers deep under the oceans.

general theory which explains the patterns of the earth's volcanic activity and earthquakes, and helps to understand the forces that produce them. According to this theory, which is known variously as "plate tectonics," "continental drift," or "sea-floor spreading," the outer surface of the earth is composed of from 20 to 30 giant slabs (or plates) which are moving about at rates of a few centimeters a year, carrying the continents and oceans with them. A typical plate may be thousands of kilometers across but only 75 to 100 kilometers thick.

The geographical patterns of earthquakes and volcanic eruptions indicate that they are concentrated at the boundaries between plates, where the plates are either separating or colliding together. Where two plates are moving apart, as at the Mid-Atlantic ocean ridge, new molten material rises from the earth's mantle, and masses of lava are added to the edges of the plates. Collisions between plates produce major earthquake activity and may raise up high mountains. The Andes are rising and producing active volcanoes because the Pacific Ocean plate is slowly shoving its way under the South American continent. The Himalayas are the result of a continuing collision between the plate which carries India and the larger Eurasian plate on which China is located. The great scar of the San Andreas Fault and the disastrous California earthquakes are produced as the Pacific plate slides northward along the western edge of the North American continent.

The pattern of the plate motions is much clearer than our understanding of what *causes* the plates to move. We know that beneath the solid plates there is a zone of weak and partly molten rock that lies between about 75 and 250 kilometers below the surface. Here, not as far from the surface as New York is from Philadelphia, the temperature is about 1,500° C. and the pressure is about 30,000 times that at the surface of the earth. This weak zone forms a kind of lubricating layer along which the plates slide and in which molten lava is generated to feed surface volcanoes. However, the forces that pull the plates along are still not well understood. The plates may be driven by movements of material in the deeper mantle below the zone of weakness. Another possible explanation is that when the edge of a plate begins to descend under a second plate, the force of gravity continues to pull the rest of the descending plate downward. There are a variety of theories in which plates are pushed, pulled, or carried apart, and the nature of the moving forces is one of the biggest unsolved problems of the earth.

A century of study, however, has succeeded in deciphering the nature of the earth's interior and understanding the basic personality of the earth. The earth, we have learned, is a hot, active planet with a strong magnetic field. There is continuous seismic and volcanic activity going on in its outer 200 kilometers. Since our planet formed, its surface has been continually reshaped; the ocean basins, which now cover 71 percent of its surface, did not exist 200 million years ago when the current episode of plate movements began.

Thanks to our new knowledge of the earth's interior, and the instruments which helped us to acquire it, we were ready to explore the interior of a second planetary body; and just as soon as the Apollo program gave us access to the lunar surface, the exploration of the moon's interior began.

MOONQUAKES: THE PULSE OF THE MOON

The internal energy of the earth (or any other planet) makes its way to the surface in a variety of ways. A lot of this energy reaches the surface either in the steady outward flow of heat or in the more irregular eruptions of molten lava from volcanoes. However, an important component of the earth's energy is slowly stored up in "strained" rocks within the earth, and then suddenly and unpredictably released in the form of earthquakes.

The slow but irresistible collisions of the moving plates that make up the surface of the earth cause large masses of rock to bend and deform within the earth. "Strain energy" slowly builds up in these deformed rock masses in much the same way as it builds up in a tree limb that is bending with the weight of a child in a swing. The strain may accumulate for decades or centuries, but eventually the rock mass reaches a breaking point; it snaps, and the stored energy is released as violent movements that spread out through the earth.

Such sudden breaks often occur repeatedly along great fractures, called *faults*, in the earth's crust, some of which, like the well-known San Andreas Fault in California, may extend for hundreds of miles. Close to the fault, the ground may suddenly move as much as several meters, causing total destruction of buildings, railroads, and highways. Farther away from the fault, the ground motion decreases, but the waves produced by a large earthquake can still be detected by sensitive instruments thousands of kilometers away.

The amount of energy released by an earthquake depends on the volume of strained rock involved. At the breaking point, each cubic meter of rock near the fault has accumulated about as much energy as that released by a small firecracker. But a large earthquake may involve millions of cubic kilometers of the earth's crust, releasing a total energy equal to the detonation of many hydrogen bombs.

The catastrophic results of earthquakes are well known. Over 830,000 people were killed in an earthquake in Shen-Shu, China, in 1556,* and the San Francisco earthquake of

* A modern repetition of this disaster occurred on July 28, 1976, when two major earthquakes struck the region around Peking. No casualty figures have been released by the People's Republic of China, but it is estimated that at least a hundred thousand and possibly half a

1906, which only killed 500 people, is very much in our minds at present because of the possibility of a future disastrous earthquake along the still-active San Andreas Fault. But these destructive events have also provided scientists with an essential tool for mapping the interior of the earth. The development of instruments called *seismometers*, which can detect ground motions as small as a millionth of a millimeter, has made it possible to record and analyze earthquake waves even after they have traveled for thousands of kilometers through the earth's interior. By recording earthquake waves, scientists can determine the location and depth of the earthquake, the amount of energy released,* and the nature of the rocks through which the earthquake waves have traveled. By examining tens of thousands of earthquake records collected for nearly a century, scientists have slowly constructed the present picture of the earth's interior.

---

million people were killed. Eyewitness accounts by foreign visitors report that the cities of Peking, Tangshan, and Tientsin (total population more than eight million) were heavily damaged.

* The energies of earthquakes are commonly measured on the Richter Scale of magnitude, named after the American seismologist who developed it. The basic equation is:

$$\log E = 5.8 + 2.4\ M$$

where *log E* is the logarithm of the earthquake energy in ergs and *M* is the magnitude on the Richter Scale. An increase in magnitude of 1 unit corresponds to an increase of about 250 times in the amount of energy released.

A sensitive seismometer can detect earthquakes of about magnitude $M = 1$. The energy of such a tiny earthquake is approximately equal to that released by a small firecracker, and the resulting ground motion would be undetectable by a human being. About 800,000 small natural earthquakes of about $M = 2.0$–3.4, which go unnoticed by human beings, are detected every year. Earthquakes of higher magnitudes ($M = 3.5$–4.5) can be felt by human witnesses. Serious damage begins at about $M = 5.5$ and earthquakes with magnitudes above about $M = 8$ usually cause total destruction in the immediate vicinity. For example, the 1971 San Fernando, California, earthquake had a magnitude of 6.6, the 1906 San Francisco earthquake was 8.25, and the 1964 Alaska earthquake was 8.6,—one of the largest ever measured. No earthquake of magnitude 9 has ever been recorded.

The analysis of moonquakes was one of the few methods to provide detailed information about the internal structure of the moon and the possible presence of either intense volcanic activity or plate motions. The lunar seismometers were the most modern products of decades of instrument development. On the moon, they could measure motions of a ten-millionth of a millimeter, detecting tiny quakes on the opposite side of the moon. Computers were used to analyze the flood of returned data, producing in seconds results that once would have taken days to resolve. And because almost nothing was known about the moon's interior before the first seismometers were placed on its surface, any results would be a major step forward.

The first lunar seismometer, placed on the moon by the Apollo 11 astronauts during man's first moonwalk, operated for 21 days* and detected only a few moonquakes. The results told scientists that the interior of the moon was much quieter than the interior of the earth; a similar seismometer on the earth would have recorded about 300 quakes during the same 21-day period.

The seismometers on the moon detected several different kinds of lunar events. They recorded quakes generated deep inside the moon. They also detected tremors produced by the impacts of small meteorites onto the lunar surface. Be-

* The Apollo 11 seismometer was built to operate as a single, self-contained instrument. It was therefore equipped with solar panels that converted sunlight into electricity, and the power supply did not survive the first lunar night. Later seismometers were all part of a group of scientific instruments (called ALSEP for Apollo Lunar Surface Experiment Package) powered by a central unit that produced electricity from the radioactive heat of a small amount of plutonium. These later seismometers, emplaced by the Apollo 12, 14, 15, and 16 missions, have been amazingly successful and have operated for periods far exceeding their expected one- to two-year lifetimes. The Apollo 12 seismometer has operated continuously since November, 1969. The complete network of four seismometers, which makes possible more precise detection and location of moonquakes, has been operating since April, 1972.

cause the inside of the moon is so quiet, and because meteorites hit the airless moon at their original cosmic speed, the impact of a meteorite weighing more than 10 kilograms (about the size of a large grapefruit) striking anywhere on the moon can be detected. About 100 such impacts are recorded every year. Fortunately, meteorite impacts and moonquakes produce slightly different records, so that it is possible to tell them apart.

However, simply detecting a moonquake is only the first step. To use the data to understand the lunar interior, we must determine when and where the quake occurred, and how much energy it released. Even on the earth, where there are hundreds of operating seismometers, this information is sometimes difficult to obtain from earthquake records. On the moon, using only a few instruments, the difficulties are that much more formidable, and the deciphering of the moon's internal structure would have been a long and arduous process if we had been limited to natural lunar events.

Fortunately, there was a way to produce well-defined artificial quakes on the moon. The Apollo spacecraft system consisted of several parts, or stages, that were crashed into the moon as they were no longer needed. These crashes produced artificial moonquakes whose locations, timing, and energies were so accurately known that the seismometer records of these events could be used as standards for interpreting the records of natural moonquakes and meteorite impacts.

The two parts of the Apollo spacecraft system that were available to make artificial moonquakes were the Lunar Module that landed on the moon, and the third stage of the series of rockets that launched the Apollo spacecraft from earth. (The larger first and second stages fell back to earth after their fuel was exhausted.) The third stage, called the S4B, carried the spacecraft into orbit around the earth and was then used to propel it toward the moon.

These artificial impacts were large enough to echo through the whole interior of the moon. An S4B stage weighs about 14 tons and is traveling at about 2½ kilometers per second when it hits the moon; the crash releases about as much energy as 45 tons of TNT. The smaller Lunar Module, weighing only 2.3 tons, crashes at about 1.7 kilometers per second, releasing the energy of about 3 tons of TNT. The lunar seismometers easily detected these impacts, even at distances of over 1,000 kilometers away.

On November 20, 1969, the newly installed Apollo 12 seismometer detected the first artificial moonquake as the discarded Apollo 12 Lunar Module was crashed into the moon near its original landing site in Oceanus Procellarum. As the record of this impact appeared on the instruments back on earth, there was sudden surprise among the watching scientists, because the record was totally unlike that of any terrestrial earthquake.

The first indication of the impact was a sharp pulse recorded by the seismometer. On earth, this vibration would have ended after only a few seconds or minutes as the earthquake waves were gradually absorbed by the earth's crust. But instead of dying away, the lunar signal continued to build to a crescendo before very slowly fading away (*Figure H*). "The moon rang like a bell," said one scientist, and it continued to reverberate for over two hours.

As more records were obtained, scientists began to see why the moon behaved so strangely. They concluded that the outermost few kilometers of the moon must be composed of cracked and shattered rock which permits the signals to echo back and forth for hours. Furthermore, this outer zone must be completely dry, because any water present would have absorbed the signals very quickly, as it does in wet soils and water-bearing rocks on earth.

In the course of the Apollo Program, nine such artificial impacts were produced on the moon. The data they provided made it possible to check the operation of the lunar seismometers, to interpret natural moonquakes more accurately, and to map the interior of the moon. There are few cases in which discarded machinery has been put to such good scientific use.

The measurement of natural internal moonquakes also pointed to fundamental differences between the earth and its moon. Compared to the earth, the interior of the moon is quiet and inert. The moon has none of the "background noise" that is produced here by minor earth tremors, volcanic eruptions, ocean surf, or trucks passing on nearby highways. Because the moon is so quiet, the seismometers could be operated at maximum sensitivity, detecting tiny

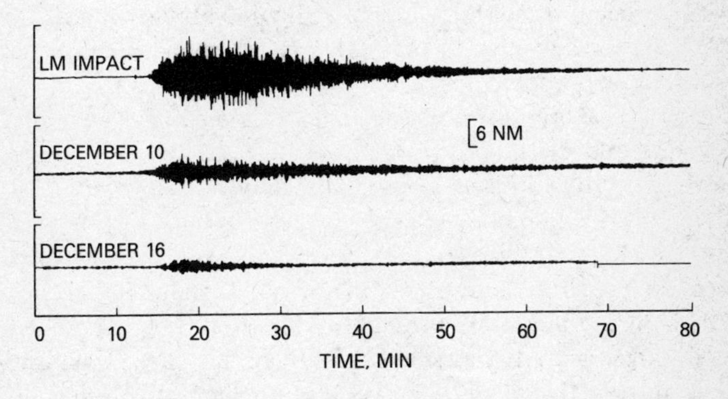

Figure H / A Moonquake Writes its Autograph. The signatures of three separate moonquakes appear in these records obtained from the Apollo 12 seismometer during the latter part of 1969. Time (in minutes) is plotted horizontally. All three signals begin at the left-hand side of the record, grow to a peak, and then gradually decline over a period of more than an hour. The vertical scale bar at right, labelled "6 nm," indicates the height of the seismometer signal produced by a ground movement of only one ten-millionth of an inch. (Diagram courtesy of G. V. Latham and G. Simmons, from *On the Moon with Apollo 15*, by G. Simmons, NASA Educational Publication, [Washington, D.C.: 1971] U.S. Government Printing Office.)

quakes that would go unnoticed on the earth. Operating a seismometer on the moon is like having an instrument in New York that can detect a sledgehammer hitting a rock in Philadelphia.

Yet even at maximum sensitivity, the lunar seismometers detect only about 3,000 moonquakes a year, nearly all below Magnitude 2. The earth generates about 800,000 quakes of this size every year. The total energy released by moonquakes is about one million-millionth of the earthquake energy released by the earth, and a moderate-size Fourth of July fireworks display gives off more energy in a single evening than the whole moon releases in a year.

The records of thousands of moonquakes have made it possible to establish the basic patterns of the moon's internal activity. First, moonquakes occur very deep in the interior; they all seem to originate in a zone about 600 to 800 kilometers below the surface. Most quakes on earth occur less than 200 kilometers below the surface, and only a few take place at depths as great as 600 kilometers. Thus, our considerable earthquake activity occurs at shallow depths, while the much weaker moonquake activity is concentrated deeper in the moon.

In addition to being deep, most moonquakes are also localized; that is, they occur again and again at specific places in the moon. At present, about 40 such "centers" of moonquake activity have been identified, all but one on the near side of the moon. The centers are not distributed at random. With only a few exceptions, they lie along two great belts that are 100 to 300 kilometers wide and run for about 2,000 kilometers across the moon. These centers, and the belts they form, may be regions where the moon's interior is slowly but continuously deforming. There may be active fractures within the moon at these points, perhaps accompanied by small amounts of molten rock.

Another unusual feature of moonquakes is their timing: they occur at regular intervals. On the earth, quakes seem to occur at random times, but on the moon, concentrations of moonquakes are detected about every 14 days—a period which corresponds roughly to half the time that it takes the moon to go once around the earth. More detailed studies have shown that moonquake centers produce a series of quakes about every 28 days, or once every lunar month. Nearly half of the centers are active when the moon is at one point in its orbit around the earth, and the other locations become active when the moon reaches the opposite point in its orbit, about two weeks later. The combination of these two 28-day cycles produces a 14-day cycle in moonquake activity for the whole moon.

The regularity of the quakes as the moon reaches certain points in its orbit around the earth suggests that moonquake activity is triggered by tidal forces generated in the moon by the earth's gravity. Tidal effects produced on the earth by the moon are well known; visible tides of several meters are produced in the earth's oceans, and smaller, but still measurable, tides are produced in the earth's rocks (see pp. 51–52). The stronger gravity of the earth results in tidal forces on the moon that are six times as large as those produced by the moon on the earth. Such tidal forces are not strong enough to break solid rock in the lunar interior, but they may add enough strain to already fractured rocks within the moon to move them, causing a moonquake—much like "the straw that broke the camel's back."

The discovery that moonquakes are related to tides may be important to understanding our own planet. It has also been suggested, though never proven, that earthquakes are triggered in a similar way by tidal forces. Further study of the moon will help us understand the tidal triggering mechanism so that we can determine what importance it has on the earth.

However, not all moonquakes are regular or related to tides. The lunar seismometers have also detected another kind of moonquake that occurs in groups at apparently random times, apparently unrelated to the moon's motions around the earth. These moonquakes are similar to the swarms of small, shallow earthquakes that often accompany the eruption of volcanic lavas on the earth. This type of poorly understood moonquake may be related to the movements of still-molten rock within the moon.

INTERNAL STRUCTURE: THE LAYERS OF THE MOON

Apollo studies have firmly established the basic nature of the interior of the moon. The moon is not a uniform body; it is made up of a series of layers that in some ways resemble the layers that make up the earth (*Figure I*). The lunar seismometers, assisted by the artificial impacts, have given us a fairly good picture of the lunar interior to depths of a few hundred kilometers. Below this level, the picture becomes less clear, for the data depend on uncertain moonquakes and unpredictable meteorite impacts. The nature of the deep interior of the moon, from about 1,000 kilometers down to the center of the moon, at a depth of 1,738 kilometers, is almost unknown. It is still an open question whether or not this region contains a small metallic core.

The outer part of the moon is best known in the region around Oceanus Procellarum, where the four seismometers from the Apollo 12, 14, 15, and 16 missions are located. There, immediately beneath the thin layer of lunar soil, is a zone of broken and cracked rock about a kilometer thick. This zone of shattered bedrock is probably the result of deep fracturing produced by large meteorite impacts, and it is this broken layer that is responsible for much of the long and unusual reverberations of moonquake echoes.

Underneath this broken material is a thicker layer of more solid rock that extends about 25 kilometers down. Because the properties of this rock are so similar to those of

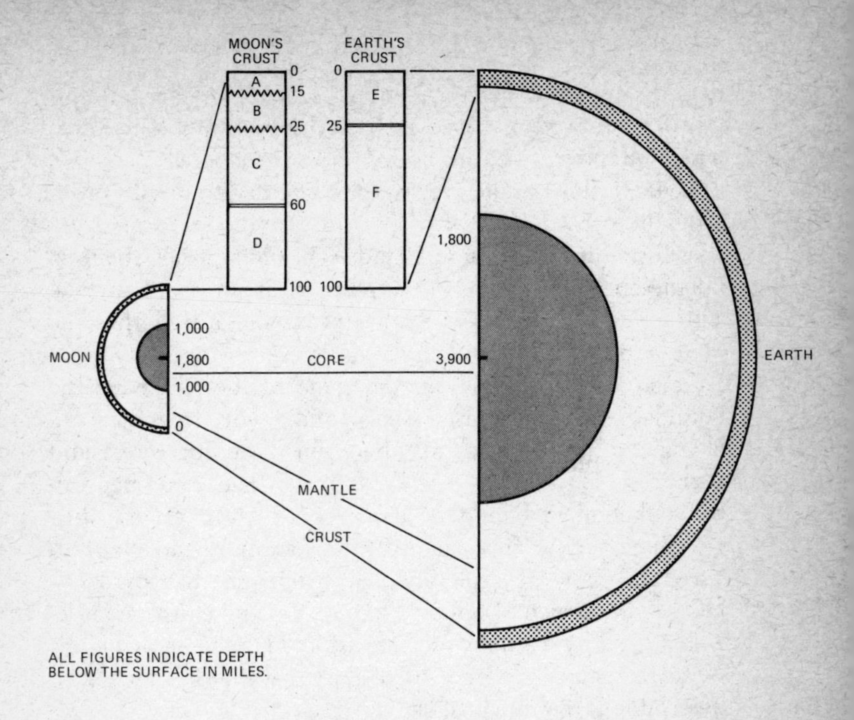

Figure I / Inside the Earth and Moon. The inside of the moon, as deciphered from the region around Oceanus Procellarum (where our instruments have been placed) is compared with the inside of the earth. The upper layer of the moon (the crust) seems thicker and more rigid than the crust of the earth, and the nature of the material beneath the lunar crust is still not well known. It is also not clear yet whether the moon has a small iron core; by contrast, the iron core of the earth is as large as the whole moon.

In this diagram, the earth and moon are drawn to scale, with the moon about one-quarter as large as the earth. Both the earth and moon are composed of three layers: an outer thin *crust*, an inner *mantle*, and an inner *core* (the existence of a core in the moon is not yet certain). The different units that make up the crusts of the earth and moon are shown, enlarged, in the columns at the top. The moon's crust consists of (1) an outer layer of shattered and broken rock about a mile thick; (2) a layer of basaltic rock about 25 miles thick; (3) a layer of feldspar-rich rock about 35 miles thick. The earth, by contrast, has a crust of mixed granitic and basaltic rock which is about 25 miles thick. On both the earth and moon, the outer crust covers a layer of denser rocks that form the mantle.

basalt samples collected from Oceanus Procellarum, this layer is believed to be a stack of the type of basalt flows which fill the mare basins in this part of the moon.

Beneath this basalt layer is a layer of material with properties like those of the feldspar-rich gabbros and anorthosites found in the lunar highlands. This layer, itself about 35 kilometers thick, extends to a depth of about 60 kilometers. This layer is probably the lunar crust which underlies the mare basins and their filling of basalt.

At 60 kilometers, still another kind of rock appears; the boundary at this point is sharp and seems to indicate a change as significant as that between the earth's crust and its underlying mantle. The lunar rocks that occur beneath the 60-kilometer boundary are not yet well identified; they are much denser than the overlying gabbros and anorthosites, and they are probably rich in minerals like pyroxene and olivine which also make up much of the earth's mantle. This dense layer extends to at least 150 kilometers, which is as far down as the properties of the lunar interior have been determined from the artificial impacts.

Below 150 kilometers, the lunar interior, whatever its composition, seems hard and rigid to depths of 600 to 800 kilometers, where the occurrence of moonquakes indicates that deformation, possibly accompanied by volcanic activity, is still going on.

The nature of the central part of the moon was partly revealed by the fortunate impact of a 1-ton meteorite which struck the far side of the moon on July 17, 1972. The vibrations from this impact passed through the center of the moon before being detected by the seismometers on the near side. As the waves passed through the center, they were altered in much the same way that earthquake waves are altered when they pass through zones of partly molten rock inside the earth. This suggests that there is a zone 1,000 to 1,200 kilometers deep in the moon where temperatures may be high enough so that the rocks are partly molten.

The question of whether the moon has a central iron core like the earth's is still unanswered. Current data do not prove that one exists, but a small metallic core 1,000 to 1,500 kilometers in diameter could be present without being detected. Such a core would be much smaller in proportion to the earth's core and would make up only about one-twentieth of the volume of the moon.

The composition of such a core is equally uncertain. A pure iron core could not be any larger than 1,000 kilometers in diameter without producing a detectable effect on the moon's orbital motions. Furthermore, because of the high melting point of iron (1,535° C.), it would be difficult both to form such a core and keep it molten. If the core contained some iron sulfide (FeS) along with the iron, the melting point would be lowered considerably (to about 1,000° C.); in this case, the moon might contain a partly molten metallic core about 1,400 kilometers in diameter. Such a core would be large enough to have produced the magnetic field detected in the returned lunar samples. At present, however, the deep lunar interior, core or no core, is still largely unexplored. We can only hope for a few more lucky meteorite impacts in the right places to help settle the questions.

Despite all the data collected by the seismometers, the internal structure of the moon has been explored in only one small area on the near (or earth-facing) side of the moon. One of the most basic questions about the moon still remains largely unanswered: Are there major differences in the internal structures of the moon's near-side and far-side regions?

Obvious surface differences between the two sides of the moon have been observed ever since the first far-side photographs were obtained by unmanned spacecraft in the early 1960s. Large impact craters, and even larger circular basins, occur on both sides of the moon in equal numbers, but the basalt flows that form the dark maria are found almost entirely on the near side. On the far side, there are only a few scattered patches of dark mare material, one of which fills the striking crater Tsiolkovsky (photo 31). Because the mare lavas were generated deep inside the moon before rising to the surface, the uneven distribution of lavas on the moon's surface suggests major differences between the two sides in the make-up of their interior.

Because the Apollo Program was restricted to landings on the moon's near side, surface instruments have never been placed on the far side. However, the scientific lunar orbit experiments provided clues about the moon's interior by carefully measuring its shape and noting the variations in its gravitational pull. Some of the most useful data came from the laser altimeter, a device that uses a laser beam to measure surface elevations with an accuracy of 1 meter (see p. 111).

The data from the laser altimeter confirmed, with far more precision, the pre-Apollo observation that the moon is not quite a perfect sphere (*Figure J*). It is very slightly elliptical, with a longer dimension about 2 kilometers greater than its mean diameter of 3,476 kilometers. The longer diameter points toward the earth.

The laser altimeter also reported that the distribution of mass within the moon is not uniform. The center of the moon's mass is offset from its geometric center by about 2 kilometers, again in the direction of the earth (see *Figure J*). These distortions of the moon's shape in the direction of the earth may have been created by strong tides when the moon was young and much closer to the earth. The bulges may also make it possible for the earth's gravity to hold the moon in such a way as to keep its near side always turned to the earth.

These variations in the lunar interior also imply that the lunar crust under the far-side highlands must be thicker than on the near side. On the near side, the boundary between the lunar crust and the denser rocks below occurs at a

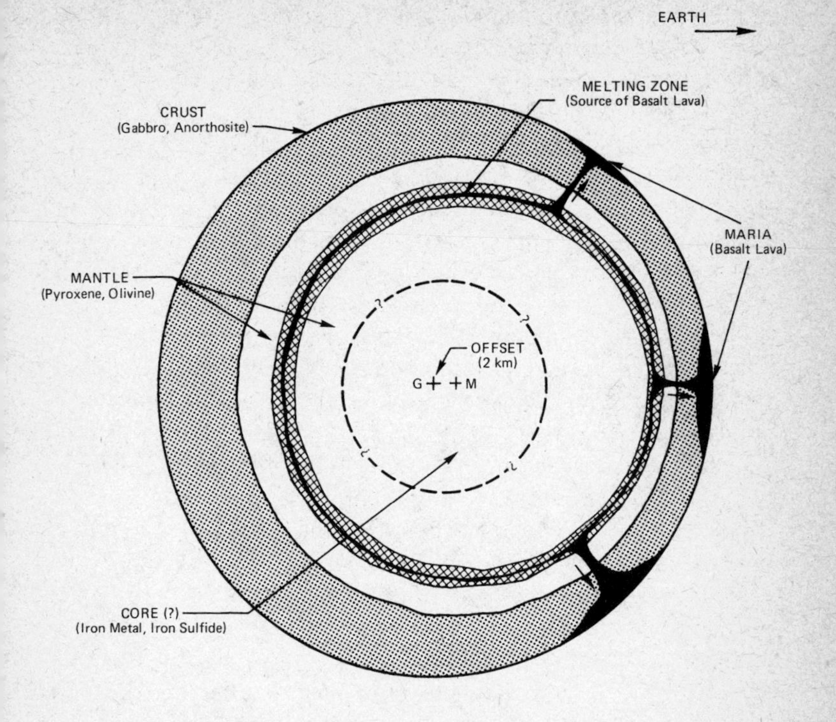

Figure J / A Slice through the Moon. The internal structure of the moon, as determined by the Apollo Program, is shown in this cross section. The moon's diameter is about 3,500 kilometers, and the different layers are not drawn to scale. The outer crust (dotted) is thicker on the far side of the moon (about 100 kilometers) than it is on the near side (about 60 kilometers). This crust is rich in calcium and aluminum and is composed of such rocks as gabbro and anorthosite. Beneath the crust is a denser mantle (white), rich in magnesium and probably composed mostly of the minerals pyroxene and olivine. A small iron-rich core (dashed boundary) may exist at the center of the moon. The moon's center of mass (M) is offset about two kilometers toward the earth from its geometric center (G). The maria (black) on the near side are filled with basalt that formed in a deep zone of melting within the moon's mantle and then rose to the surface (arrows).

depth of about 60 kilometers, but the orbital data suggest that the same boundary must occur at a greater depth—about 75 to 100 kilometers—on the far side.

The thicker crust on the far side of the moon might explain why there is so little lava on that side. A thicker crust would likely contain less radioactive elements, thus making it difficult to heat the lunar interior to the melting point in order to form lavas on the far side. Furthermore, it might make it more difficult for any lava formed beneath it to reach the surface. On the other hand, if these theories are correct, it is hard to explain why Tsiolkovsky, a crater only 200 kilometers in diameter, should be filled with mare basalt while much larger, deeper, and older impact basins remain empty.

We cannot answer these questions fully with our current data. A complete answer will probably require the establishment of a seismometer network on the far side to determine the structure there, combined perhaps with a landing in Tsiolkovsky to sample the dark material and determine exactly what it is and when it formed.

The great thickness of solid rock that forms the outer part of the moon demonstrates a fundamental difference between the moon and the earth. On the earth, weak and partly molten rock forms a layer only about 75 to 250 kilometers deep. The rock that makes up the moon is solid to a depth of at least 600 to 800 kilometers. The upper part of the moon is too cold and rigid to develop a system of moving crustal plates like those on the earth, and the present styles of deformation of the two planets are totally different.

If there is any molten rock inside the moon, it must occur below the thick outer part, in and below the zone of moonquakes. There is some evidence that small amounts of molten material exist at depths of about 1,000 to 1,400 kilometers inside the moon. Surface heat flow measurements and calculations based on the electrical properties of lunar rocks suggest that temperatures inside the moon could be as high as 600 to 800° C. at depths of about 200 kilometers and could reach the melting points of basalt and other lunar rocks (about 1,200° C.) at depths of about 1,000 kilometers. This is just about where the occurrence of random moonquakes, together with the changes in moonquake waves, have indicated the existence of some molten rock.

The existence of molten rock deep within the moon may explain the puzzling "lunar transient events" that have been reported by astronomers for hundreds of years. Sir William Herschel (1738–1822) reported observing bright red glows on the moon in 1783 and 1787 and suggested that they were lunar volcanic eruptions. Subsequent observers have reported about 1,200 events—such as glows, hazes, brief color changes, and temporary obscurations of lunar surface features. Many of the observations, made in recent times by careful observers, are convincing, although no adequate explanation has yet been proposed, and no transient events have been observed by astronauts.

The transient events seem restricted to certain places on the moon. Certain craters have been the site for many events; about 300 have been reported from the crater Aristarchus, 75 from Plato, and 25 from Alphonsus, including Kozyrev's 1958 detection of a possible lunar gas emission (see p. 61). Other transient events seem to be concentrated around the edges of the maria—the same regions in which the detection of radon emitted from the lunar interior has also been reported (see pp. 114–15).

The cause of lunar transient events is still unknown. Present-day volcanic eruptions are hard to reconcile with the ancient ages of all the lunar lavas. Other explanations, like electrically generated lunar dust clouds, or fluorescence of a weak lunar atmosphere by solar radiations, are hard to prove. The possible existence of molten rock deep within the moon makes it easier to argue that the transient events are caused by small amounts of volcanic gas, or a mixture of dust and gas, rising to the surface through fractures in the crust. If this hypothesis is correct, then future observations should show a close relation between transient events and the timing of moonquakes, and transient events would then be the visual result of deformation inside the moon.

A landing and a sample return from some active spot like Aristarchus may be needed to settle the origin of these transient events. They may eventually secure us proof that the deep interior of the moon is hot and partly molten. Until they are explained, they serve as a reminder that although the moon has been cold and quiet for the last 3 billion years, it may not be completely dead.

MAGNETISM: WHERE DOES A LUNAR COMPASS POINT?

Explorers and other travelers have known for hundreds of years that the earth has a strong magnetic field that causes compass needles to point north. In more recent times, the magnetic properties of the earth have been used for navigation, prospecting for metal deposits, measuring the ages of rocks and human artifacts, and establishing the slow drifting of the continents away from each other.

Our ability to apply the earth's magnetism has always been far ahead of our understanding of how it originates. In 1600, Sir William Gilbert (1540–1603) suggested that the magnetic field existed because the whole earth behaved like a permanent magnet.* However, later discoveries about the

* A *permanent magnet* is a piece of metal which produces a magnetic field because its atoms are all lined up in the same direction. The common household or laboratory magnets, made out of alloys of iron, nickel, or cobalt, are permanent magnets.

A magnetic field can also be produced by the passage of charged particles through a conductor. *Electromagnets* are made by passing an electric current through windings of copper wire; in these magnets, the magnetic field exists only when the current is flowing. A magnetic earth indicated that its interior was too hot for any material to act as a permanent magnet. According to more recent theories, the earth's interior is a kind of electromagnet or dynamo in which the earth's metallic core plays an essential role.

The formation of the earth's core and the resulting development of its magnetic field was a major event in the earth's history. At the time that the earth formed, more than a third of its weight was made up of iron. In the original earth, this iron may have been scattered through the whole planet, but apparently the earth soon became hot enough to melt. This melting was severe enough to change the iron into droplets of molten metal which slowly sank through the less dense rocks to form a central core. This poorly understood process of heating and separation was so effective that it formed an iron core nearly half the diameter and one-eighth the volume of the earth itself.

As far as we can tell (and we cannot tell very well), the earth's magnetic field is produced by electric currents that circulate through the core of the earth just as they do through a generator or an electromagnet. A more detailed explanation has been hard to develop. For one thing, we know very little about the physical, chemical, and magnetic properties of the solid and liquid metals in the core, where temperatures reach several thousand degrees Centigrade and pressures reach hundreds of thousands of atmospheres.

Another problem with developing an explanation is the strange behavior of the magnetic field itself. The magnetic poles wander slowly over the earth's surface, and every million years or so the magnetic field suddenly reverses itself, so that a compass needle would point south instead of north. The last change, from a "reversed" period of time to

---

field can also be produced by charged particles moving in a vacuum; a weak but detectable magnetic field is produced by the charged atoms in the solar wind.

our present "normal" period, occurred only 700,000 years ago, during the lifetime of some of man's direct ancestors. These reversals, which have been traced back for several hundred million years, have provided an important new tool for determining the ages of rocks, but they are difficult to reconcile with modern theories about the origin of the magnetic field itself.

The discovery of a magnetic field around a planet is thus an important key to its history and internal structure, for it implies the presence of an iron core which melted and separated itself within the planet during an early period of intense heating. Since a magnetic field can be detected at a significant distance by a passing space probe carrying sensitive detectors, we do not have to land on a planet to determine its magnetism. During the last decade, unmanned spacecraft have measured the magnetic fields of four other planets. The probes have yielded a wide variety of results, but little logic, concerning planetary magnetism.

Mars, a planet about half the size of the Earth, has no detectable field. However, Mercury, which is even smaller than Mars, has a definite field that was detected on the Mariner 10 flyby in March, 1974. Mercury's field is about 200 gammas*—not even 1 percent of the Earth's field, but large enough to indicate that the planet has a significant iron core. Venus, a larger planet often described as "Earth's Twin," has no detectable magnetic field at all. Among the planets so far studied, Jupiter has the largest magnetic field.

* The basic unit for measuring the intensity of a magnetic field is the *oersted* (oe) named for the Danish physicist Hans Christian Oersted (1777–1851) who discovered electromagnetism. The magnetic field of the earth is about 0.6 oersted at the poles and about 0.3 oersted at the equator. A laboratory magnet can produce much stronger fields—10,000 oersteds and up. Natural magnetic fields are conveniently described by a smaller unit called the *gamma,* which is equal to one hundred-thousandth oersted. The earth's magnetic field, which is fairly large as planetary magnetic fields go, is 30,000 to 60,000 gammas.

Close-up measurements made on the Pioneer 10 flyby in December, 1973, indicate that Jupiter's magnetic field is 10 to 15 times as strong as the Earth's. This intense magnetic field may be related somehow to the bursts of radio wave "noise" that have been detected from Jupiter for many years.

Before the Apollo landings it was generally believed that the moon had either a very weak magnetic field or none at all. Unmanned spacecraft had detected fields of only a few gammas near the moon, and even this weak field was probably associated with the solar wind. The moon, it appeared, had no iron core and no magnetic field.

One of the major surprises to come out of the Apollo Program was the discovery that the moon had apparently possessed a fairly strong magnetic field at some time in the past. The evidence came from the magnetic properties of returned lunar rocks, from measurements made by astronauts on the lunar surface, and from instruments on small satellites orbiting the moon.

The unusually strong magnetism in lunar rocks was first observed in the Apollo 11 samples and has been observed in samples from every subsequent mission. The magnetism is the stable "remanent" type that develops in terrestrial lavas and other rocks when they cool in the earth's magnetic field. These results indicated that a strong magnetic field had existed on the moon when the various lavas and breccias had cooled between 4.2 and 3.2 billion years ago. The magnetic field of the moon had been impressed on the tiny particles of iron metal in the cooling rocks.

Careful studies indicated that this remanent magnetism had developed on the moon and was not a result of exposing the samples to the spacecraft or to the earth's magnetic field. To check this possibility fully, a sample of an Apollo 12 basalt (sample 12002) was completely demagnetized on earth and then sent back to the moon with the Apollo 16 mission to see what magnetism it would acquire during the two-way trip. Although the sample picked up some weak magnetism, none of the stronger "remanent" magnetism was detected when the sample was returned to earth and re-measured.

In addition, the fact that astronauts could actually measure significant magnetic fields on the lunar surface itself was strong evidence that the magnetism was a strictly lunar product. At the Apollo 12 landing site in Oceanus Procellarum a value of 38 gammas was measured, and a lower value of 6 gammas was measured by the Apollo 15 astronauts at Hadley Rille. These lower values seem typical of mare regions, while stronger magnetism was found at the highland sites. At Fra Mauro (Apollo 14) local fields of 43 and 103 gammas were measured. Even higher values were found at Descartes (Apollo 16); a field of 121 gammas was measured on the traverse, and the field of 313 gammas recorded at North Ray Crater was the strongest measured on the lunar surface. These high magnetic fields seem to be associated with the Cayley Formation, one of the units of breccia exposed at the Apollo 16 site.

The measurements by the astronauts showed that significant local magnetic fields are associated with the rocks that form the lunar surface. This conclusion was supported by magnetic measurements made from moon-orbiting satellites. During both the Apollo 15 and 16 missions, "subsatellites" launched from the Command Module measured the magnetic field around the moon for several months after the astronauts had returned to earth. A third satellite, Explorer 35, gained additional data on its elongated orbit around the moon. Although Explorer 35 was originally launched to study the effects of the earth's magnetic field on the solar wind, it made unexpected and important contributions to mapping the magnetic fields on the moon itself.

These satellite measurements made it possible to construct a fairly comprehensive magnetic map that showed areas of strong and variously oriented magnetism tens to hundreds of kilometers in size. The lunar surface seems to be made up of variously magnetized blocks; there is no single north or south magnetic pole on the moon.

In general, the maria on the near side of the moon have lower magnetism than the highlands—less than 50 gammas—and are more uniform magnetically. The highland areas, especially those on the far side, show stronger, if uneven and variable, magnetic fields which are often above 300 gammas. A strong magnetic anomaly occurs just north of the crater Van de Graff on the far side. Van de Graff is a unique area in other respects as well: it forms a depression about four kilometers deep in the otherwise high far side of the moon, and it also shows slightly higher radioactivity than its surroundings. This anomaly may indicate buried volcanic rock beneath the region; in any case, it marks Van de Graff as a spot that merits further study.

The evidence from samples, astronauts, and satellites shows that a significant magnetic field must have existed on the moon between about 4.2 and 3.2 billion years ago. It is possible to estimate the intensity of this ancient lunar field from the magnetism still remaining in lunar samples, but the procedures are uncertain and complicated. Estimates for the lunar field range from about 1,000 gammas (a few percent of the earth's field) to 120,000 gammas (about twice the value of the earth's present field). Obviously, no one is too sure how intense the lunar field might have been or how much it might have varied in the past.

The evidence for ancient magnetism on the moon is probably the most puzzling result of the entire Apollo Program. It has generated a host of possible explanations, none of which is very satisfactory. Perhaps the most logical suggestion is that the moon, like the earth, had a metal core which operated to produce a magnetic field and then somehow switched off sometime during the last three billion years. But it is not easy to explain how this core could produce a strong magnetic field for over a billion years and then cease to operate. A large body like the moon loses internal heat only via the slow conduction through its interior. If the moon had a molten, magnetic core 3½ billion years ago, the core should still be molten and the moon should still have a magnetic field. If the deep lunar interior is still hot, as a lot of our evidence indicates, then the absence of a lunar magnetic field is even harder to explain.

Another explanation is that the lunar magnetism was acquired from the earth's magnetic field at a time when the moon was much closer to the earth, perhaps as close as 20,000 kilometers. But although the moon could have been this close to the earth soon after it formed, there seems to be no way the earth could have kept it that close for the time span (over a billion years) required to magnetize both the old and young lunar rocks. More exotic explanations involving unusually strong magnetic fields in the sun or in the solar wind entail the same problem. There is no evidence that such abnormal fields ever existed, and it is hard to find a mechanism that would provide such fields for a billion years or more after the sun had apparently settled down to its present form of "normal" behavior.

Basically, we just do not know enough yet to understand where the lunar magnetism came from. We do not know yet whether the iron-bearing lunar basalts have different magnetic properties from the iron-free terrestrial rocks with which we are more familiar. We do not know if, or how, the magnetic properties of lunar surface rocks might have been altered by the shock waves produced by meteorite impact, or by the bombardment of higher-energy atomic particles from the sun. We do not even understand the generation of the earth's magnetic field well enough to interpret what the lunar magnetism may tell us about our own planet. The lunar compass is pointing into the past, but we are not yet wise enough to read the directions.

CHAPTER

THE MOON AND BEYOND

In the centuries before the Apollo Program, we watched the moon as we might watch a stranger passing to and fro outside our house. Now we have gone outside to meet the stranger. The moon has become an acquaintance (photo 32), and she has now revealed to us much of her own personal history.

The illumination of the moon's past is probably the greatest scientific triumph of the Apollo Program, for we now have the record of another world to compare with the history of our own planet.

The preserved lunar record is more ancient than our own. (See Table 2) The lunar rocks tell us that the moon formed as a separate body about 4.6 billion years ago in the course of the overall formation of the solar system. Like the other planets, the moon is probably an accumulation of many smaller bodies that coalesced to become one large body.

At first, the moon was a molten, battered world. The formation of the moon via collisions with small objects re-

Table 2 / The Life Histories of the Earth and Moon. This time chart sums up the great differences in the histories of the earth and moon. The moon preserves a record of its early events and shows practically no activity more recent than three billion years ago. By contrast, the earth's early history is no longer preserved, and most of the earth's events are relatively recent, including the development of life in the oceans, the appearance of land mammals, and the rise of man.

MOON	MILLIONS OF YEARS AGO		EARTH
Formation	4,600		Formation
Intense melting, highlands formed, probably continuous impacts	4,400		
Oldest preserved rocks	4,200		
Great impacts	4,000 to 3,800		
		3,750	Oldest preserved rocks (Greenland granites)
Filling of maria by lava flows	3,700 to 3,300	3,400	Oldest volcanic rocks, oldest traces of life, record of oceans
		2,600	Maximum age of most oldest rocks
Youngest volcanism(?)	2,100		
		2,000	Two large impact craters (30–60 kilometers diameter)
Copernicus crater formed by impact	1,000		
		600	Development of higher life
Tycho crater formed by impact	300	300	Land animals and plants
		200	Most recent separation of continents
		100	Dinosaurs
		50	Formation of Rocky Mountains; mammals
		3	Man

leased so much energy that the outer part of the moon was completely molten to a depth of several hundred kilometers. Certain minerals then separated out from this moon-wide ocean of melted rock, producing the chemical differences that are reflected in the present-day lunar geology. As the moon slowly solidified, continuing collisions with large bodies broke up its ancient rocks and also pitted the surface of the lunar highlands with craters. This intense bombardment ended about four billion years ago with a series of huge impacts that excavated the great mare basins like Mare Imbrium and Mare Orientale.

Then, internal radioactive heating began to melt the moon again, this time in a relatively narrow zone between 100 and 250 kilometers beneath the surface, causing great eruptions of lava that spread out over the lunar maria for more than half a billion years, from about 3.8 to 3.1 billion years ago. After the last of these eruptions about three billion years ago, the moon became quiet. Meteorite impacts have sculpted its surface, building up the soil layer and sometimes forming large craters like Copernicus and Tycho. Except for the tremors of moonquakes and the still unexplained "transient events" (see p. 237), the moon is inactive now.

Despite the flood of chemical and historical information obtained by the Apollo Program, we still do not have a single, universally accepted theory for the origin of the moon. Because scientific theories always die hard, the three pre-Apollo theories (double planet, fission, and capture) have all survived the Apollo results, though often with considerable modifications.

A completely successful theory of lunar origin must explain the evidence that the moon has been a separate and unique body since its formation. It also must account for significant differences in the chemistries of the earth and the moon. This chemical disparity is the major stumbling block of the "double planet" theory, which argues that the earth and moon were both formed together in the collapsing dust cloud that became the solar system. It is hard to believe that such major chemical differences could have been produced in two bodies that formed so close together. Consequently, most current explanations for the origin of the moon combine modifications of the other two traditional theories, fission and capture.

But the original version of the fission theory—that the moon spun off as a single body from a rapidly rotating earth—has also been undercut by the Apollo data. The chemical differences between the earth and moon, especially the absence of volatiles in the moon, are so profound that it is hard to argue that the earth and moon ever were part of the same body.

A newer variation of the fission theory suggests that the moon was built up gradually from a heated atmosphere that was thrown off a hot, rapidly spinning primordial earth. During its formation, the earth was heated up by collisions with small bodies until the temperature in its outer layers was over 2,000° C. At such a high temperature, both volatile materials and some less volatile elements like silicon, aluminum, and magnesium boiled off the primitive earth into a dense atmosphere around it. As this atmosphere cooled, the less volatile elements condensed into small rocky particles which were spun into orbit around the earth and then assembled to become the moon. The moon thus developed by separating from the earth atom by atom instead of by separation as a single mass. Since it holds that the moon formed from material whose volatile elements were removed by the intense heating, this theory does account for the different chemical compositions of the earth and moon; but, like the original fission theory, it still has a number of problems.

Apollo's confirmation of the chemical differences between the earth and moon has led still other scientists to argue that the moon formed somewhere else in the solar dust cloud and was then "captured" by the earth. However there is some disagreement as to *where* in the solar system the moon might have formed. The loss of volatile elements indicates a high temperature of formation, which prompts some scientists to place the origin of the moon near the center of the solar system, inside the present orbit of Mercury. But if it had originated there, it would have developed an enrichment in iron, as the dense planet Mercury apparently did. Unfortunately for this argument, the moon has a relatively low iron content. Another snag in the capture theory is that the captured body has to be slowed down in order to go into orbit around a planet like the earth. A possible explanation is that the moon was slowed down by crashing into a swarm of smaller bodies which circled the earth at that time.

All of these theories explain some of the data about the moon, and all of them run into difficulties trying to explain all of it; none of them can be conclusively proven or disproven. We will probably never understand the origin of the moon until we make progress in understanding the formation of the solar system itself. We have a great deal more to learn about the actual chemical processes that went on in the original dust cloud. We also need to know more about the mechanical processes which caused small particles to assemble into larger bodies and then brought these bodies together into moons and planets. When we understand these mechanisms better, we may be able to put more precise boundaries on where the moon actually originated. If it is proven that the moon originated inside the orbit of Mercury, then some kind of capture process must have occurred, no matter how unlikely it may seem. On the other hand, if new theories manage to explain how chemically different bodies could form close together, then the "double planet" origin may be correct after all.

New questions are constantly arising to complicate any single explanation of the solar system. If the formation of large moons was a normal phenomenon when our solar system formed, where are the large moons that we would expect to circle Mercury, Venus, and Mars? It could be that tidal forces on the sun have destroyed any original moons of Mercury and Venus. However, this explanation will not work for Mars, which has only two tiny captured asteroids instead of a full-fledged moon, because it is farther from the sun than the earth.

Another important post-Apollo question is whether the moon is chemically unique. Six large moons, about the same size as ours, circle the giant planets of our solar system—four around Jupiter, one around Saturn, and one around Neptune. We do not know yet whether these other moons share the high-temperature history and other chemical peculiarities of our own moon, or whether they are mostly condensed ice like the planets they accompany. An unmanned space mission to Jupiter's moons could answer this question.

THE MOON AND THE EARTH

What we have learned about the moon has also revamped our thinking about the earth. Although the earth and moon have different chemical compositions and different histories, the moon is still an important model of what the primitive earth may have been like. The moon clearly records a primordial melting and widespread chemical separation that produced a layered internal structure almost immediately after it had formed. It is likely that the present internal structure of the earth, including its iron core, also developed very early in its history, perhaps as a result of the accretion process that formed it.

The intense early bombardment recorded by the moon more than 4 billion years ago may be a general characteristic of the solar system too. If a similar intense bombardment struck the earth at this time, it would help explain why no terrestrial rocks older than 4 billion years have been found.

The discovery of ancient rocks on the moon has also generated a new enthusiasm for probing the ancient history of our own planet. Some of this excitement derives from the discovery of rocks about 3.8 billion years old in Greenland. These unusually old terrestrial rocks were found at about the same time that the Apollo 11 mission was collecting lavas of the same age from Mare Tranquillitatis. Moreover, the surprising discovery that the lunar highlands are composed of plagioclase-rich rocks such as anorthosites and gabbros (see pp. 133–34) promptly spurred a renewed interest in a group of similar terrestrial rocks which occur only in minor amounts in geologically old regions. The origin of these terrestrial anorthosites is an old unresolved geological problem. Comparative studies of terrestrial and lunar anorthosites may help explain the origin of both types of rocks as well as explaining why a rock that is a minor curiosity on earth is one of the fundamental building blocks of the moon.

The earth and moon provide two contrasting examples of how differently planets can develop, and in their contrast we can see some of the factors that control the evolution of planets. Size is important. A large planet can hold volatile materials like water, and it can also retain more internal heat to produce continuous geological changes. Chemical composition is also important; a planet without water, no matter how large, lacks the one substance that is essential for the only kind of life we know. The presence or absence of radioactive elements determines whether a planet will be hot or cold during its lifetime, and the amount of iron in a planet determines whether it can ever develop a strong magnetic field. The first two bodies we have explored, the earth and its moon, show two different lines of development. Although we think that the planets all formed in the same general way, it is almost certain that we will find further different planetary histories as we explore the solar system.

AND THE LANDS BEYOND

The planets have suddenly become familiar too, for the manned exploration of the moon has gone hand in hand with the unmanned exploration of the solar system. In the five years between 1968 and 1973, scientists launched 17 heavily instrumented spacecraft to investigate every one of the five planets known to ancient astronomers. Mariner 9 went into orbit around Mars in November, 1971, circling the planet like a tiny third moon and radioing back to earth over 7,000 pictures of craters, volcanoes, canyons, and sand dunes on the Martian surface. Two years later, in December, 1973, Pioneer 10 passed safely through the Asteroid Belt, carried its instruments in a sharp turn around the giant planet Jupiter, and then sped away to become the first man-made object to leave the solar system. A sister spacecraft, Pioneer 11, made the same trip successfully a year later, swinging around Jupiter in December, 1974, and then heading outward on a path that will bring its instruments and cameras close to the ringed planet Saturn in September, 1979.

Mariner 10 was launched in the other direction, inward toward the sun. It passed close to Venus in February, 1974, and then settled into an orbit around the sun so well planned that the spacecraft has already been able to make three close approaches to the planet Mercury, taking pictures of that planet that are as good as our best views of the moon through earthbound telescopes.

We can now apply the lessons from the moon to other planets. We have learned, for example, that the early period of intense bombardment and cratering observed on the moon seems also to have been general throughout at least the inner part of the solar system. Mercury has revealed a battered surface that is virtually identical to the lunar highlands (photo 33). The surface of Venus is shrouded in clouds, but earth-based radar, probing through its atmosphere, has detected a number of large circular depressions that are almost certainly craters. Mars exhibits two kinds of terrain. Half its surface is heavily cratered, while the other half is covered with younger features that seem to be volcanic lavas, wind-blown dust, and possible river channels.

Furthermore, the processes of chemical separation, melting, and volcanism also seem to have occurred on the other planets. Mercury has a detectable magnetic field, indicating the existence of an iron core. Photographs of the surface show units that resemble volcanic deposits. However, the moon has taught us to be cautious about interpreting photographs too quickly. There was a time before the Apollo 16 landing when most scientists thought that the Cayley Formation was volcanic too, and it turned out to be composed of impact-produced breccias.

The occurrence of chemical separation on Venus has been demonstrated only by a single analysis of the surface made by a Russian lander, Venera 8, in July, 1972. The lander survived for nearly an hour on the surface and sent back an analysis resembling the composition of granite, a rock that requires considerable chemical processing to produce, at least on earth.

Mars provides unquestionable evidence of chemical evolution and volcanism. The lightly cratered half of the planet contains numerous structures that are undoubtedly volcanoes. They resemble the volcanic peaks of the Hawaiian Islands, but on a much greater scale. Mars' largest volcano, Olympus Mons, is 600 kilometers in diameter—about the size of the State of Nebraska—and rises about 25 kilometers above the surface of the planet (photo 34).

Not only have we discovered that the moon is not a primordial object, but our exploration of the solar system has already shown that none of the small planets is likely to be an unaltered sample of the solar system. Perhaps we will have to search for original solar system material in Jupiter, in its icy moons, or in the bodies of comets that occasionally pass by us. We may even discover that the evolution of the solar system has so altered all the matter in it that there is nothing left of the original ingredients.

Our view of the distant universe also changed radically during the decade we spent exploring the moon. The same technology that carried man to the moon also invented new optical and radio telescopes to study the far corners of the universe from the surface of the earth. From new observations, and from unexpected discoveries of strange objects like quasars and pulsars, astronomers have recently been able to put boundaries on the size and age of the known universe. The universe seems to have an edge, or a boundary, at the unimaginable distance of about 15 billion light-years* away from us. Ages calculated for the universe fall in the range of 10 to 15 billion years; it is somewhat surprising that the universe seems to be only three or four times as old as the earth.

Coincidently, there are three general theories to explain the origin of the universe, just as there have been three main theories to explain the origin of the moon. The "Big Bang" theory suggests that the universe began at a single point in an inconceivable explosion of matter and energy

* A *light-year* is the distance that a beam of light, with a velocity of 300,000 kilometers a second, travels in a year. One light-year is equal to 9.47 million million kilometers, or about 6 million times the distance of the earth from the sun. The nearest star is about 4 light-years away, and our own Milky Way Galaxy is about 75,000 light-years in diameter.

and that the parts of the universe have been flying apart ever since. A second model, the "Steady State Universe," holds that matter and energy are being continually created and destroyed, with no beginning or end. And a third idea is that we live in an "Oscillating Universe" that first expands from a primordial explosion, then slows down and stops, and finally collapses back on itself to start the whole process again.

The discoveries of the last few years have strengthened the "Big Bang" theory. The explosion apparently occurred 10 to 15 billion years ago, and the "edge" of the universe that we can barely see is the outer shell that has been expanding at about the speed of light ever since the explosion occurred.

There is no better testimonial to science and technology than the fact that we have been able to put limits on what was once considered to be an infinite and eternal universe. But we are not yet close to a final answer. Even the supporters of the "Big Bang" theory are divided as to whether the universe will continue to expand forever or whether it will collapse again to make another cosmic fireball in another 15 or 20 billion years. Furthermore, the mere existence of limits immediately raises questions about what lies beyond them.

Perhaps in our universe, time, space, and matter are so closely connected that none of them can be considered separately. But the simple fact that so many new questions can be raised and discussed is a measure of our tremendous progress. About 200 years ago, a pioneer geologist named James Hutton (1726–1797) commented that the earth showed "no vestige of a beginning, and no prospect of an end." It took man about 150 years to put boundaries on the life span of the earth, and only a few decades more to extend the boundaries to the universe itself.

NEW MOON, NEW MEN

The Apollo Program reaffirmed two of man's basic traits: curiosity and tool-making. The astronauts, engineers, and scientists are not far removed in spirit from the men who chipped flint axeheads, who recorded their world in the cave paintings, and who raised the great boulders to make the observatory at Stonehenge. Apollo is only another episode in the story of man's use of his mind to understand his surroundings and his increasing ability to build the machines that his mind conceives.

However, the Apollo Program also inaugurated a new dimension in man's history. Man is now a space traveler. He can leave the earth, he can live and work in space, he can explore other planets. Space has suddenly become a permanent part of the human environment, just as the dry land long ago became a new world for man's remote ancestors.

Furthermore, space is rapidly becoming more familiar as we continue to explore it. We now know that space travel, while complicated and expensive, is neither impossible nor particularly dangerous. A Mars-bound astronaut would be much safer than the early explorers who crossed the Atlantic in tiny wooden ships to explore America, knowing far less about the dangers of land and ocean than we now know about the hazards of space.

By going to the moon, we have also obtained a new view of the earth, just as anyone develops new insights about his native land by traveling abroad. From the moon, the earth is a beautiful blue globe in a black, impersonal sky (photo 35). It shows no national boundaries, no signs of strife, and no evidence of the thin film of human civilization on its surface. The Apollo Program has given visible meaning to the term "Spaceship Earth," and as we continue to explore space, we must also turn our new skills back toward our own planet in order to learn how it operates and how we can continue to survive on it. The ancient, sterile desert of the moon is a grim reminder how narrow the line is between life on earth and the lifelessness of our surroundings.

ARE WE ALONE IN THE UNIVERSE?

The moon has no life, but even its sterilized soil contains traces of the chemicals out of which life can be built. We have learned that the basic ingredients for life are common: a sun, a few ordinary atoms, and some interstellar dust. Then, in a protective incubator like the earth, simple molecules can change and become more complex, finally crossing that uncertain line that separates chemical molecules from living things.

As we learn more about the universe, we realize that these conditions must have been achieved thousands, or even millions, of times. In our own modest Milky Way Galaxy, there are at least 100 billion stars. Many of these stars are like our sun, many of these suns must have planets, and many of these planets must be like the earth. The conditions that produced life on earth no longer seem unusual. Unless there is some incredible combination of improbabilities that has made us unique, it is likely that the story of life has been repeated in the universe many times with many variations.

The question of life elsewhere could also be answered for us in a manner over which we have no control. Before we expect it, other life may come to search for us, for we have been loudly proclaiming our existence to the universe for over half a century. Since about 1900, our radio, TV, and radar signals have been spreading outward from the earth like electromagnetic ripples, carrying the proof of our existence out into the universe at the speed of light, 9½ million million kilometers every year. These signals have already passed nearly a hundred stars, including such familiar tenants of our own night sky as Sirius, Procyon, Vega, Castor, and Pollux, and the signals will still be recognizable when they have passed many thousands more. If other inhabitants of the universe have developed intelligence, curiosity, and technology, they may come to see for themselves what these strange signals mean.

Whether life exists elsewhere or not, man will never be entirely of the earth again (photo 36). The sky above us is no longer a barrier but an endless ocean, and we stand on the shore, carefully testing the water. From our own planet, the moon, the other planets, and even the stars stretch out like a series of unknown islands, encouraging us to explore further. We have made our first brief, cautious voyage. The flags and machines that we have left on the moon will last for millions of years. Either they will be seen again by other men, or they will announce our existence and our accomplishment to other space travelers whose form we cannot even imagine.

Like all great events in history, Apollo is a turning point, another fork in a long road. One road leads outward to further exploration and to new knowledge. The other road leads back to Earth and to ourselves, to a more intelligent operation of our planet. The lesson from Apollo is that neither road by itself is enough. From now on we must walk both roads.

WHAT NOW?

On December 14, 1972, the Apollo 17 Lunar Module blasted off from the Littrow Valley, ending the series of Apollo lunar landings and carrying home the eleventh and twelfth men who had walked on the surface of the moon.

Apollo 17 is not the end of man's exploration of the moon.

After all great breakthroughs, whether they occur in art, in science, or in exploration, there has always been a period of consolidation and preparation for the next great leap. If the history of our race tells us anything, it tells us that where man has once stepped, he will walk again.

The same technology and desire for knowledge that sent us to the moon is now reaching out a thousand times further to begin the close-up exploration of the distant planets. On July 20, 1976, exactly seven years after Neil Armstrong set foot on the moon, a robot spacecraft streaked through the thin air of Mars, slowed in a sudden flare of rockets, and settled gently on a red and rock-strewn plain in a region called Chryse. With this safe touchdown, the Lander half of a two-part spacecraft called Viking 1 successfully completed a one-year, 640-million-kilometer journey from the surface of earth to the surface of Mars. Overhead, the Orbiter half of Viking 1 circled Mars to relay back to earth the Lander's discoveries about what Mars was like.

Viking 1 was the fourth attempt to soft-land a spacecraft on the surface of Mars. In 1971 and 1973 the Russians landed three small spacecraft (Mars 2, Mars 3, and Mars 6), but all three failed during or shortly after landing. Viking's success was partly due to the fact that the spacecraft could make a detailed survey of Mars to find the safest landing site before the Lander was sent down.

After arriving at Mars, Viking 1 remained in orbit for a full month while scientists on earth feverishly studied pictures returned by the cameras in the Orbiter, discarding possible landing sites that the new, sharp pictures showed to be rough and hazardous. At the same time, the large radio telescope at Arecibo, Puerto Rico, bounced radar waves off Mars, measuring the roughness of other landing sites from more than 320 million kilometers away. With dozens of photographs and hundreds of radar returns, the safe landing of Viking 1 was made possible, and it was followed on September 3, 1976, by the equally successful descent of the

Viking 2 Lander to a more northerly region of Mars called the Plains of Utopia.

The Viking Landers are an almost miraculous technological achievement. Instruments that would fill whole rooms have been redesigned to fit into a small spacecraft that is less than 3 meters across and weighs only 605 kilograms. The Lander looks like a cluttered hexagonal table on three legs, but it contains a power station, a television studio, a weather station, an earthquake detector, a miniature backhoe for sample collecting, two chemical laboratories (one for inorganic and one for organic analyses), and three different incubators for growing and detecting any life that might be present in the reddish soil of Mars.

The pictures of the Martian surface taken by the Lander's cameras are clear and sharp, despite being radioed back more than 320 million kilometers (a distance so great that, even at the speed of light, the radio waves take twenty minutes to make the trip). The pictures show a level plain, covered with fine material and strewn with volcanic-looking rocks. The "soil" is like the lunar soil or like wet sand on earth: fine-grained, cohesive, and strong enough to support the Lander. In the distance are small dunes, accumulations of wind-blown red dust on and behind boulders, and patches of what may be exposed bedrock.

An unprotected man could not survive on the surface of Mars. A blast of lethal ultraviolet radiation from the sun strikes the Martian surface, and there is too little air to breathe. The atmosphere is less than 1 percent as dense as earth's, and measurements made by the Landers show that it is mostly carbon dioxide (95 percent), with minor oxygen (0.1–0.4 percent), nitrogen (2–3 percent) and argon (1–2 percent).

The weather on Mars is sunny, cold, and stable. A typical Mars weather report for the Viking 1 Lander's location reads: "Clear and cold. Minimum temperature − 122°F. (− 86°C.); maximum temperature − 25°F. (− 31°C.).

Pressure 7.7 millibars (about 0.008 atmosphere). Light winds from the east in late afternoon, changing to southwest after midnight. Average wind velocity 2.4 meters per second (5 miles per hour), gusting to 10 meters per second (22 miles per hour)." Since Viking 1 landed, the atmospheric pressure has been slowly dropping, possibly because carbon dioxide in the atmosphere is freezing out onto the polar ice caps as winter comes to the southern hemisphere of Mars. The Martian sky is not blue like earth's, but is a pale pink, because the atmosphere carries a great amount of fine red dust.

Eight Martian "days" after Viking 1 landed, a small scoop stretched out from the spacecraft, dug into the soil nearby, and transferred soil samples to several different instruments. The chemical composition of the soil, measured by bombarding it with X-rays, turned out to be somewhat like basalt, supporting the idea that the bubble-rich rocks around the Lander may be volcanic lavas. The elements aluminum, silicon, sulfur, calcium, titanium, and iron have been detected in the Martian soil. The large amount of iron (14 percent) supports the view that the red dust of Mars is an iron oxide or hydroxide not unlike terrestrial rust.

If the rocks on Mars are lavas, then they seem different from both terrestrial and lunar lavas. The soil contains less aluminum than do terrestrial basalts, and it has much less titanium (less than 1 percent) than do lunar lavas. The high sulfur content (2–5 percent) in the Martian soil is unusual and may be produced by oxidized sulfate minerals. Furthermore, some water (less than 1 percent) is released from the soil when it is heated. Conditions on Mars are clearly different from those on the moon. There is water and oxygen on Mars, and the soil seems chemically weathered in ways that resemble the weathering and oxidation that occur on earth.

The most ambitious task of the Viking Landers was to detect any forms of life that might exist, even dormant, in the Martian soil. Three separate instruments served as incubators for soil samples, mixing them with water and nutrient solutions, then analyzing to detect any substances like carbon dioxide or oxygen, that might be given off by living organisms.

There was a burst of excitement when strong activity was detected in the Martian soil samples, but this excitement has been mixed with a great deal of caution. No one is yet certain whether the reactions that have been observed are produced by living things or by strange chemical reactions between the Martian soil and the water and nutrients provided by the instruments. It is puzzling to find the Martian soil reacting when other Viking instruments have detected in it none of the carbon-bearing organic chemicals that are fundamental to all forms of terrestrial life. Until the experiments are completed and the data are analyzed fully, the scientists are justifiably cautious. It is still too soon for Viking to answer the question, "Are we alone"? It is almost unbelievable that we have even been able to ask this question on the surface of yet another world.

In many ways Viking has brought Mars as close as the Surveyor landings brought the moon only a decade ago. We now know that Mars, like the moon, has a surface that will support machines and men. We see rocks that can be collected. We have our first rough chemical analyses of the soil. We know the surface conditions, and we can build equipment to meet them. With what we have learned from the Apollo Program, we can follow Viking with robot sample-return missions or with men.

Viking is not the only voyage, nor is Mars the only destination. Other ambitious missions being planned for the future depend on the positions of the planets for their timing. Two Mariners, 11 and 12, are slated for launch in 1977, to reach Jupiter in 1979 and to pass by Saturn in 1981. Two Pioneers are being prepared for a 1978 launch toward Venus, each one a spacecraft that will orbit the planet and drop probes through the thick, boiling atmosphere to the surface.

The United States has not been alone in the exploration of the planets. Seventeen out of the thirty-one launches between 1961 and 1976 were Russian spacecraft aimed at Venus and Mars. Several actually landed on the surface of each planet, and a few survived long enough to return significant information. Venera 8, which in 1972 returned the first chemical analysis of the surface of Venus, was actually the sixth Russian lander on that planet; its data suggested a rock composition like that of terrestrial granite. In the Fall of 1975, two more Russian landers, Venera 9 and 10, landed safely on Venus and transmitted back the first photographs of the planet's surface. The pictures showed huge rock slabs and boulders exposed to the planet's dense, scorching atmosphere.

The long-separate Russian and American space programs came together briefly in the Apollo–Soyuz Test Project in July, 1975, when three astronauts in an Apollo Command Module linked up with two Russian cosmonauts in a Soyuz spacecraft. This "handshake in space" lasted only a few days, but it was the result of several years of cooperative preparation in which Americans and Russians visited each other's launching sites and worked with each other's equipment. The Apollo–Soyuz effort is a hopeful sign that the two great competitors in space can sometimes put aside their differences long enough to become collaborators.

APOLLO: THE UNKNOWN LEGACY

We have only begun to investigate the scientific results of the Apollo Program, and even our study of the lunar samples themselves is just getting started. The six Apollo landings returned a total of 382 kilograms (842.3 pounds) of lunar rock and soil, and this total now includes over 50,000 separate samples—large and small rocks, complex breccias, small fragments from the soil, and 1- and 2-meter cores from the lunar soil layer. Despite the intense efforts of hundreds of scientists for more than six years, only about one-quarter of this material has been examined, and detailed analyses have been made on only about one-tenth of it.

Everything we know about moon rocks has been learned from only a small fraction of the returned material. The preservation and careful study of the remaining lunar samples are an essential part of the future exploration of the moon, and NASA is continuing an active program of lunar sample studies. The most common lunar rock types have now been well described, and scientists can now search the remaining rocks and breccias for unusual fragments. This research might discover some unexpected varieties of lunar rocks, or rocks that have come from deep within the moon, or lavas that are younger than any so far analyzed.

The soils, the breccias, and the core samples are being given special attention because they contain a record of the history of the sun, the nature of cosmic radiation, and the presence of volatile materials. There is enough material here to keep several platoons of scientists occupied; one gram of lunar soil may contain as many as ten million separate tiny particles, each of which may have an individual history and a unique record of the solar wind. The opening and study of the deep core tubes, some of which contain half a billion years of lunar history, will be a major scientific project by itself.

BACK TO THE TELESCOPE

The Apollo Program spawned a new era in earth-based lunar astronomy. For the first time in history, man can now focus his telescopes on another world whose surface composition has been determined from returned samples and whose surface features have been observed at close range. Using the standards provided by the Apollo Program, astronomers can now determine the chemical composition, the topography, the heat flow, and the nature of the surface layer over the entire earth-facing side of the moon.

Today's astronomers, thanks also to their modern instruments, are no longer limited to making visual descriptions of the moon. Analyses of the sunlight reflected by the lunar soil can provide information about the chemical composition of the soil. The weak infrared and radio waves emitted by the moon contain information about the lunar surface temperature and heat flow. Radar beams from earth bounced off the moon by instruments powerful enough to reach the planet Saturn make it possible to explore the shape of the lunar surface and to trace out its mountains and valleys within an accuracy of a few meters.

ROBOT ASTRONAUTS OF THE FUTURE

The next machine to go to the moon will probably be a Lunar Polar Orbiter, a spacecraft designed to observe and analyze the whole surface of the moon from lunar orbit. It would be placed so that its orbit passes over the north and south poles of the moon instead of being limited to the region around the lunar equator analyzed by the Apollo missions.

The Lunar Polar Orbiter will make it possible to extend the scientific measurements that so far have been made over only about 20 percent of the moon's surface. In a polar orbit around the moon, the spacecraft would eventually pass over its entire surface, because the moon rotates on its axis once every month while the Orbiter is passing over it.

Lunar samples have already been obtained by mechanical means. On September 20, 1970, an unmanned Russian spacecraft called Luna 16 landed in Mare Fecunditatis. Using a hollow drill, the spacecraft collected a 100-gram (3½-ounce) sample of lunar soil and returned it to earth.

In an important step in international cooperation in space, some of the Luna 16 material was given to American scientists in exchange for Apollo 11 and 12 soil samples.

The instruments available for analysis of lunar samples are so sensitive and so precise that the small amount of Luna 16 material obtained from the Russians (about 3 grams—the weight of ten aspirin tablets) yielded an impressive amount of information. Scientists discovered that the surface of Mare Fecunditatis was covered by titanium-poor basalt lavas about 3.4 billion years old. The lavas were similar to, but slightly older than, the rocks returned from Oceanus Procellarum (Apollo 12). Even at a considerable distance from the Apollo landing sites, the lavas still seem to be about 3½ billion years old. The Luna 16 soil had also been heavily exposed to cosmic and solar atomic particles while on the moon, but it is hard to interpret the effects because there is so little material available for study.

Seventeen months later, on February 21, 1971, Luna 20 landed in an area of the lunar highlands between Mare Fecunditatis and Mare Crisium. Samples of this soil, also exchanged with American scientists, proved to be made up of crushed plagioclase-rich rocks very similar to the breccias returned by the Apollo 16 mission from the highlands near Descartes.

The Russians are still continuing their sampling of the moon by robot spacecraft. On August 18, 1976, Luna 24 landed safely in Mare Crisium, a small circular mare on the eastern edge of the moon. The spacecraft drilled about 2 meters deep into the lunar soil, then returned safely with a core section of the soil layer that provides a unique sample of the nature and history of an unknown part of the moon.

Existing spacecraft or their more complex descendants, can easily return similar samples from the moon. Our analytical instruments make it possible to obtain a great deal of scientific information from a tiny sample, and our experience with the larger samples obtained by the Apollo missions provides a necessary check on our interpretations.*

In future missions, it will be possible to combine unmanned sample collection with a roving vehicle that moves over the lunar surface. So far, the Russians have the only experience with unmanned lunar roving vehicles. Some of their Luna missions, instead of returning samples, landed a wheeled vehicle called *Lunokhod* (roughly, "moonwalker"). Controlled from earth, the vehicle traveled over the lunar surface, transmitting back TV pictures and data about the physical and chemical nature of the surface.

Future roving vehicles may travel for hundreds of kilometers, measuring the chemistry, gravity, and magnetic properties of the surface as they go. Even more complex rovers, equipped with TV cameras and guided from earth, will be able to examine the local geology and collect samples. At the end of the trip, the cargo would be transferred to a small spacecraft for return to earth.

Where should we send these machines on future missions? Landings on the near side of the moon are easiest to control because the machines are always in sight of the earth and can receive instructions continuously. The search

* There is a continuing debate over whether the scientific results from the Apollo Program could have been obtained at much less cost with unmanned samplers similar to the Luna 16 and Luna 20 spacecraft. In many cases the answer is no. The unmanned samplers returned only a small amount of soil and no large rocks; the one formation age determined for the Luna 16 material was made possible by an incredibly painstaking analytical effort on a rock chip that weighed 0.062 gram. The larger samples returned by the Apollo missions were essential to learning about formation and exposure ages, highland breccias, microcraters, lunar magnetism, the nature of solar wind, cosmic ray particles, and the layering and history of the lunar soil. The formation ages measured on large lunar rocks were especially important; without these ages, we might have concluded that the model age of the lunar soil, about 4.6 billion years, was actually the age of the mare lavas. Without the large rocks from Apollo, on which the true formation ages could be determined, our whole view of lunar history might have gotten off to a very false start, and we might never have learned that the moon had been an active evolving planet for a billion and a half years.

for young volcanic rocks in the maria is one of the most important things that could be done on the near side of the moon, because it would help establish new limits on the thermal history of the moon. Studies of the number and distribution of craters on the maria suggest that young volcanic rocks may be found in parts of Mare Imbrium and Oceanus Procellarum; ages as young as 2.5 to 1.7 billion years have been estimated for these rocks. If these ages could be verified from returned samples, the whole history of the moon would have to be revised. The Marius Hills (see photo 8), which have already been identified as some of the youngest volcanic features on the moon, are an obvious landing site for such a mission.

Another unmanned mission could try to determine the origin of lunar transient phenomena by landing where they have been most often seen, in the craters Aristarchus, Alphonsus, or Plato. In addition to collecting samples of possible recent volcanic rocks, the spacecraft could leave instruments behind to await the next "eruption"—a seismometer to detect moonquakes, a heat flow experiment, and an atmosphere detector to sample diffusing gases.

Much remains to be done on the near side of the moon, but the entire far side of the moon is practically unexplored, and scientists are eager to send instruments there and to obtain samples. The curious magnetic anomaly near the crater Van de Graaff (see p. 243) is an obvious site to place an instrument package, and the mare-filled crater Tsiolkovsky (see photo 31) could provide samples of both the highland crust and of the dark mare material that for some reason is so scarce on the far side of the moon.

Landing spacecraft on the far side of the moon is a difficult problem. On the far side of the moon, the spacecraft is out of radio communication and cannot be controlled directly from earth. One solution is to build a complex spacecraft that can be programmed in advance to land and perform its tasks without any contact with earth. However, a simpler and less expensive solution is to put a relay satellite in orbit over the far side of the moon, where it can "see" both earth and the lunar far side at the same time. Fortunately, nature has done most of the necessary work for us. Some distance beyond the moon, there is a point where the gravity fields of the earth and moon combine in such a way that a satellite placed there will always stay there, remaining on the far side of the moon as the moon orbits.

With such a relay satellite in orbit, nearly continuous communication between the earth and instruments on the lunar far side would be possible, and more ambitious explorations can be planned. The most important scientific step will probably be the landing of a group of seismometers to explore the interior of the moon beneath the highlands. More complex instrument packages could measure chemical and magnetic properties as well.

It may soon also be possible to put instruments on the far side to study things beyond the moon. The far side of the moon is an excellent place to do astronomy; it is entirely airless, utterly dark for half of the time, and shielded behind the entire mass of the moon from the lights and radio noises that make both optical and radio astronomy difficult on the earth. Small automatic telescopes placed there could make observations that are impossible for instruments on earth. They could observe the ultraviolet and infrared light of the stars; and seek new sources of radio waves, X-rays, and gamma rays in the sky; and possibly even find new examples of such puzzling objects as quasars, pulsars, and black holes.

The United States may not have to undertake these explorations alone. The new feasibility of international cooperation in exploring the moon is an important result of the Apollo Program. There is the precedent of Antarctica, where many nations, including the United States and Russia, have cooperated for over a decade in the scientific exploration of a continent about one-third the area of the moon itself. The more recent examples of the lunar sample exchanges and the joint Apollo–Soyuz mission are also encouraging indicators of future cooperation in space between the United States and Russia. It is possible now to plan a joint mission in which a Russian Luna spacecraft would descend to sample the crater Tsiolkovsky guided by an American relay satellite fixed above the far side of the moon. Geologists could easily plan voyages for a Russian Lunokhod that would collect samples along a trail a hundred kilometers long and then transfer them to an American robot spacecraft for return to earth. Or it could be an American rover and a Russian spacecraft. Neither the hazards of the moon nor the state of our technology is the factor determining whether such cooperative exploration takes place. We now know that such joint missions can be done if the governments and individuals involved decide that they should be done.

Another great gain from the Apollo Program is confidence. Already the exploration of the moon has changed from a great unknown challenge to a matter of relatively familiar engineering. What will determine the future exploration of the moon is no longer our ignorance and uncertainty about the universe, but the resources of desire, talent, and money that we ourselves provide.

In this post-Apollo period, the moon has become in some ways as familiar as the earth, and the other planets are becoming as familiar as the moon was a decade ago. No longer just our satellite, the moon has become a base and a proving ground, no longer a destination but a way station on the road of continuing exploration.

Despite our partial domestication of the moon it would be foolish to think that we have learned everything important or interesting about it. The Apollo Program has let us, as Newton put it, pick a few pebbles from the edge of a boundless ocean, but there is still much to be learned from studying the beach while we make plans to venture out onto the ocean itself.

APPENDIX 1
CHRONOLOGY OF MANNED APOLLO MISSIONS

MISSION	LAUNCH DATE	ASTRONAUTS	DURATION†
Apollo 7	October 11, 1968	Walter M. Schirra, Jr. Donn F. Eisele R. Walter Cunningham	10:20:09
Apollo 8	December 21, 1968	Frank Borman James A. Lovell, Jr. William A. Anders	6:03:01
Apollo 9	March 3, 1969	James A. McDivitt David R. Scott Russell L. Schweickart	10:01:01
Apollo 10	May 18, 1969	Thomas P. Stafford Eugene A. Cernan John W. Young	8:00:03
Apollo 11	July 16, 1969	*Neil A. Armstrong *Edwin E. Aldrin, Jr. Michael Collins	8:03:18
Apollo 12	November 14, 1969	*Alan L. Bean *Charles P. Conrad, Jr. Richard F. Gordon, Jr.	10:04:36
Apollo 13	April 11, 1970	James A. Lovell, Jr. Fred W. Haise, Jr. John L. Swigert, Jr.	5:22:55
Apollo 14	January 31, 1971	*Alan B. Shepard, Jr. *Edgar D. Mitchell Stuart A. Roosa	9:00:02
Apollo 15	July 26, 1971	*David R. Scott *James B. Irwin Alfred M. Worden	12:07:12
Apollo 16	April 16, 1972	*John W. Young *Charles M. Duke, Jr. Thomas K. Mattingly II	11:01:51
Apollo 17	December 7, 1972	*Eugene A. Cernan *Harrison H. Schmitt Ronald E. Evans	12:13:52

† Days:hours:minutes

*Asterisks indicate those astronauts who landed and walked on the moon.

APPENDIX 2
HIGHLIGHTS OF MANNED APOLLO VOYAGES

MISSION	LANDING SITE	HIGHLIGHTS
Apollo 7		First manned Apollo flight. Completed 163 revolutions around the earth.
Apollo 8		First manned voyage around moon. Completed 10 revolutions around moon. Broadcast Christmas Eve message to earth.
Apollo 9		First test of complete Apollo system in earth orbit. Separation and redocking with Lunar Module. Spacewalk by Schweickart. Completed 151 revolutions around the earth.
Apollo 10		Test of complete Apollo system in lunar orbit; "dry run" for landing mission. Completed 31 revolutions around moon. Lunar Module descended to within 15 kilometers (9 miles) of lunar surface.
Apollo 11	Mare Tranquillitatis	First lunar landing. Armstrong and Aldrin spent 21 hours, 38 minutes on moon. Returned samples of lunar soil and basalt rocks from maria.
Apollo 12	Oceanus Procellarum	Two lunar field excursions; returned samples of different younger basalt. Brought back pieces of Surveyor 3 spacecraft.
Apollo 13		Lunar landing at Fra Mauro canceled by explosion of oxygen tank in Command Module. Astronauts returned safely by using Lunar Module as a "lifeboat." Carried out lunar photography; used their S4B stage to make an impact on the moon which was detected by the Apollo 12 seismometer.
Apollo 14	Fra Mauro	Study of ancient breccias of Fra Mauro Formation. Long walk to Cone Crater (more than five kilometers).

MISSION	LANDING SITE	HIGHLIGHTS
Apollo 15	Hadley Rille and Apennine Mountains	First use of Lunar Rover. Astronauts traveled more than 24 kilometers (15 miles) on moon. Installed first heat probe experiment. Collected variety of complex rocks from canyon of Hadley Rille and front of Apennine Mountains. Extensive orbital science obtaining photographs and analyses of lunar surface. Collected "Genesis Rock" (anorthosite). Launched subsatellite into orbit around moon.
Apollo 16	Descartes Plateau	First landing in lunar highlands. Long trips with Lunar Rover to Stone Mountain and North Ray Crater. Collected shattered and deformed feldspar-rich rocks (gabbro and anorthosite).
Apollo 17	Taurus Mountains and Littrow Valley	Last Apollo mission. Long trips in Lunar Rover across Littrow Valley. Collected basalt rocks from valley floor and breccias from nearby hills. Collected Orange Soil. Installed second heat probe and verified high lunar heat flow.

BIBLIOGRAPHY

BALDWIN, R. B. *The Measure of the Moon.* Chicago: University of Chicago Press, 1963.

BARBOUR, J. (ed.). *Footprints on the Moon.* New York: Association Press, 1969.

COLLINS, M. *Carrying the Fire: An Astronaut's Journeys.* New York: Ballantine, 1975.

COOPER, H. S. F. *Apollo on the Moon.* New York: Dial, 1970.

———. *Moon Rocks.* New York: Dial, 1970.

JASTROW, R. *Red Giants and White Dwarfs.* New York: Signet, 1969.

KING, E. A. *Space Geology.* New York: John Wiley & Sons, 1976.

LANGSETH, M., and L. LANGSETH. *Apollo Moon Rocks.* New York: Coward, McCann, and Geoghean, 1972. (For very young readers.)

LEWIS, R. S. *The Voyages of Apollo.* New York: Quadrangle, 1974.

LOWMAN, P. D. *Lunar Panorama: A Photographic Guide to the Geology of the Moon.* Zurich, Switzerland: *Weltflugbild* Reinhold A. Müller, 1969.

MAILER, N. *Of a Fire on the Moon.* New York: Signet, 1970.
MUTCH, T. A. *Geology of the Moon: A Stratigraphic View.* Princeton, N.J.: Princeton University Press, 1970.
SHKLOVSKII, I. S., and C. SAGAN. *Intelligent Life in the Universe.* New York: Dell/Delta, 1968.
SHORT, N. M. *Planetary Geology.* Englewood Cliffs, N.J.: Prentice-Hall, 1975.
TAYLOR, S. R. *Lunar Science: A Post-Apollo View.* New York: Pergamon Press, 1975.
WILFORD, J. N. *We Reach the Moon:* The New York Times *Story of Man's Greatest Adventure.* New York: Bantam, 1969.

INDEX